U0329540

现代创意办公空间 II

马英伟 季雨晴 肖昕瑶 郝诗婷 范羽茗 杜虹 译

高迪国际出版有限公司 编
大连理工大学出版社

图书在版编目 (CIP) 数据

现代创意办公空间 . 2 : 英汉对照 / 高迪国际出版有限公司编 ; 马英伟等译. —大连 : 大连理工大学出版社 , 2013.2
ISBN 978-7-5611-7610-8

Ⅰ. ①现… Ⅱ. ①高… ②马… Ⅲ. ①办公建筑—建筑设计 Ⅳ. ① TU243

中国版本图书馆 CIP 数据核字 (2013) 第 020517 号

出版发行：大连理工大学出版社
（地址：大连市软件园路 80 号 邮编：116023）
印　　刷：利丰雅高印刷（深圳）有限公司
幅面尺寸：240mm × 320mm
印　　张：20
插　　页：4
出版时间：2013 年 2 月第 1 版
印刷时间：2013 年 2 月第 1 次印刷
策划编辑：袁　斌 刘　蓉
责任编辑：刘　蓉
责任校对：王丹丹
封面设计：高　迪

ISBN 978-7-5611-7610-8
定价：308.00 元

电话：0411-84708842
传真：0411-84701466
邮购：0411-84703636
E-mail:designbooks_dutp@yahoo.cn
URL:http://www.dutp.cn

如有质量问题请联系出版中心：（0411）84709246 84709043

PREFACE
序言

As more and more employees invest longer hours at work we are creating spaces that are "healthy environments" that nurture and are supportive, invigorating the minds of its employees, allowing them to feel empowered. Studies have shown that investing in life nurturing working environments affects the bottom line and makes good business sense.

Designers need to respond to the evolving workplace, to provide flexibility and mobility which encourages and supports teamwork and information exchange. We do so by taking individuals out of private spaces or by making offices smaller, adding glass to the interior walls to provide transparency, and by providing larger public, and communal spaces. The result is an interior space that fosters interaction and allows people to connect and share ideas. Wherever people work today they want to maintain contact with one another, to have the opportunity to interact in various ways, both scheduled and spontaneous, and to share ideas. Because the most important commodity is still gifted and productive talent, well designed spaces are critical for attracting this key resource and to support productivity.

Technology will allow a space to become a more integral part of the built environment, a living organism responding to peoples' needs. Systems will be aware of occupancy and changes of usage and of people. Ventilation and lighting will respond to and accommodate the number of people in a space. For security purposes, flooring will also identify who was in a space last and when. Environments will become more interactive and responsive, friendlier and seemingly more simple. The global business culture is driven by a desire for efficiency. And in Bill Gate's terms: "Innovation equals efficiency."

Environmentally responsible design will be a given. New green products are being developed at a faster rate than ever before. Changes in lighting and lighting systems and acoustics will all be integrated and more efficient.Clients' consistent concerns are delivering a functional space within budget, and within a specified timeframe, a space that fosters a positive working experience. Well designed spaces offer a more conducive atmosphere, and are attitude changing and more productive work environments. How an interior space becomes a productive resource and a healthy environment will determine the role of the designer and the direction of the workplace in the future.

Nancy Keatinge
Felderman Keatinge + Associates

由于越来越多的员工花在工作上的时间不断增加，我们正在建造一些具备“健康环境”的办公空间。这样的空间能够培养并积极鼓励员工进行发散思维，培养他们的主人翁意识。研究表明，在工作环境上的投资会影响公司的最终效益，很有商业意义。

设计师们需要顺应不断变化的工作场所，为了支持并鼓励团队协作以及信息交流，他们需要表现得灵活多变。我们通过让每名员工走出私人空间，或缩小办公室空间、在室内增加玻璃墙板来达到办公透明化的效果，并建立更大的公共社交场所。这样一来，创造出的室内空间就可以增强员工间的互动，使他们相互联系、分享彼此的想法。当今社会，不论员工们在哪工作，他们都想与同事保持联系，希望有机会用各种方式，有计划地或自发地与大家进行互动，分享意见。由于有才华并高效的人才依旧是最重要的元素，优秀的空间设计对于吸引这一关键资源、促进生产力发展至关重要。

现代技术让办公空间作为满足人们需求且充满活力的有机体，成为了建筑环境的一部分。系统可检测到使用面积和人员的变动。通风装置和照明设备会根据办公室人数的变化进行相应的调整。为安全起见，地板还可以识别何人何时最后离开办公室。办公环境将变得更具交互性和感应力，更加友好，也更加简洁。全球的商业文化受效率需求的支配。用比尔·盖茨的话说就是：“创新就是效率”。

环保设计是必不可少的。新环保产品的开发速度较以往相比快了很多。照明装置和音响装置都将变得一体化，并更高效。设计师们旨在营造一个能够培养积极的办公体验的空间，并始终考虑在预算和规定的时间范围内完成客户的功能空间设计。精心设计的空间为员工调整工作态度提供了一个更有利、更高效的工作环境。如何将室内空间设计成一个高效、健康的环境，决定着设计师的角色和未来办公空间的发展方向。

南希·基汀格

费尔得曼 & 基汀格联合公司

CON-TENTS
目录

Design Company dwp

UAWITHYA CORPORATE HEADQUARTERS

Location
Bangkok, Thailand

Area
750m^2

Uawithya is a leading name in the quarry and equipment business in Thailand. They recently commissioned world-class architecture and interior design firm dwp, to design and build their corporate headquarters on Wireless Road, in central Bangkok.

As an internationally respected company, the office was specifically designed to showcase the numerous products and services the company provides to its clients, as well as provide an efficient, comfortable working environment for its staff.

A bold statement in the design was conceptualized by using the corporate color red, as the main accent point. This was then set juxtaposing with a strong masculine palette of steel and stone, to convey the nature of the business. There is an evident air of transparency and openness throughout.

Additionally, bold graphics were utilized to create an inspirational buzz about the business within the space, and highlight the pride of the company in its past, present and future.

Uawithya
Bangkok
Saraburi
Chonburi
Uthong

Bangkok
Saraburi
Chonburi
Uthong
Surin
Udonthani
Phisanulok
Lampang
Surathani
Tungsong
Hat yai
RUSTON BUCYRUS
33-RB

Bangkok
Saraburi
Chonburi
Uthong
Surin
Udonthani
Phisanulok
Lampang
Surathani
Tungsong
Hat yai

FURUKAWA

Simple Solutions

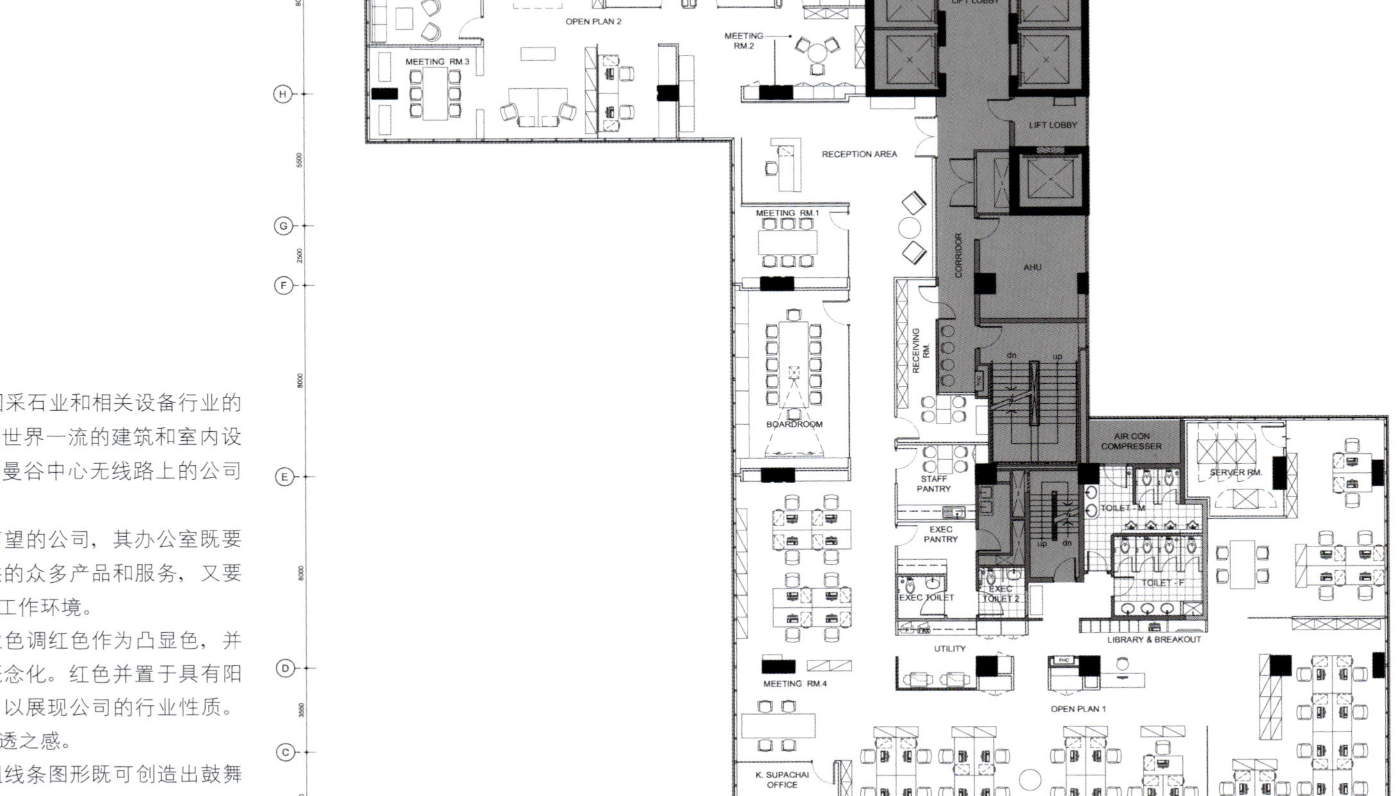

Uawithya 公司是泰国采石业和相关设备行业的领军者。最近，他们委托世界一流的建筑和室内设计公司 dwp 为其设计位于曼谷中心无线路上的公司总部。

作为一家拥有国际声望的公司，其办公室既要向客户展示公司所能提供的众多产品和服务，又要为员工提供高效、舒适的工作环境。

本案采用公司的企业色调红色作为凸显色，并使设计中的夸张表现力概念化。红色并置于具有阳刚之气的钢铁和石头旁，以展现公司的行业性质。室内整体给人以开阔与通透之感。

此外，在室内运用粗线条图形既可创造出鼓舞人心的兴奋之感，又能强调该公司过去、当下和未来的骄傲成就。

Designer Gabriel Salazar, Fernando Castañón

ACCESOLAB

Design Company
usoarquitectura

Location
Mexico City, Mexico

Area
1,000m^2

Material
Drywall, glass, millwork and carpet tiles

Photographer
Héctor Armando Herrera

Accesolab

Accesolab is a company with 20 years of experience commercializing sophisticated equipment for research laboratories. For the design of their new corporate offices in Santa Fe – a major business area in Mexico City – the team worked with usoarquitectura to develop a project to outstand their leadership.

A very simple architectonic program, adapted to the space characteristics, was set out in a 1,000 square meters area. The confidence of the clients in the project was transformed in a central square area for the operative zone with private offices located close to the facades with windows.

The services section works like a screen to increase the privacy of the direction area. This section contains the assistants, site and storage areas. Next to the building elevators, the visitor's areas were located. Training, showroom, meeting and reception halls resulted in a functional and flexible space with a vitality contagious atmosphere.

For the color palette 14 different shades of gray – both cold and warm – were selected and combined with accents in cadmium yellow, silver and white. Any suspicion caused by the selection of gray, as the central element of the palette, was minimized with the lineal pattern chosen for the carpets that elegantly combines with the woodwork in American oak colored in rat gray and plastic laminates in both gray and white.

The circle in this project is closed with the lighting design. Different temperatures of lamps were used to enhance the gray color palette. In the operative area apparent installations were used. Suspended lamps – with direct light – arranged in a linear pattern frame the random galleries that indicate the evacuation routes and add a different personality to the typical front of private offices.

Accesolab 是一家从事了 20 年实验室精密仪器生产的公司。其位于墨西哥城主要商业区圣达菲的新办公楼由乌苏建筑事务所的设计师设计完成，通过这一项目展现了设计师们的高超水平。

本案占地 1000 平方米，设计简洁又符合空间特征。令客户最满意的一点是将中央的正方形区域作为办公区，在靠近玻璃幕墙的位置设置了个人办公室。

服务区起着屏风的作用，可以很好地保护指示区的隐私。该区域包括援助站、网点和储藏区。紧挨着大楼电梯的是访客区。其中的培训厅、陈列室、会议室和接待大厅使这个实用而灵活的空间充满了活跃的气氛。

为了达到调色板的效果，设计师选取了 14 种不同色度的灰色——既有冷色也有暖色——并与镉黄、银色和白色搭配，形成对比色。灰色为本案的主色调，其冷峻的色彩感被地毯的线性图案中和到最低。地毯、美国橡木制成的鼠银色木艺品以及灰白层叠相间的塑料工艺品搭配在一起，相得益彰。

圆圈在该项目中与照明设计密切相关。应用不同亮度的灯来强调灰色的颜色搭配。办公区安装了明显的照明装置。悬挂射灯排列成线状，构成了随机的图形，这些图形不仅用来指示疏散路线，也为个人办公室特有的前门增添了个性。

Designer Hugo Ramos, Rita Pais, Jette Fyhn, Dina Castro, André Góis Fernandes, Ana Lúcia da Cruz, Ricardo Sousa, Bruno Almeida

FRAUNHOFER PORTUGAL

Design Company
Pedra Silva Architects

Project Coordination
Luis Pedra Silva with ENGEXPOR

MEP Design
JCT with GATENGEL

Lighting
Astratec with Prolicht Lighting

Location
Oporto, Portugal

Area
1,660m^2

Photographer
João Morgado

Fraunhofer Portugal is a non-profit private research association and is part of the German Fraunhofer-Gesellschaft, the largest organization for applied research in Europe. Although not familiar to the general public, Fraunhofer is responsible for many important innovations, including, for example, the MP3 file format, and many advances in workplace organization research.

Pedra Silva Architects was selected, through an open competition, to design the new Porto headquarters, located at the Technology University Campus – Parque de Ciência e Tecnologia da Universidade do Porto (UPTEC). The design took into account Fraunhofer's innovative philosophy through a message that is simple, positive and dynamic. Innovative workplace layout and organizational elements from Fraunhofer Office Innovation Center in Stuttgart (Germany) were also an important input to the project, adding another layer to the concept.

The new research facilities occupy two floors in a new UPTEC building in a total of 1,660sqm. Circulation is the project's backbone. All spaces appear along a distribution route located next to the glass façade. This main axis allows access to all different spaces. These spaces, with different functions and sizes, are generated and consolidated through a bold gesture: a waving plane that goes through the open floors, creating different spaces and ambiances.

This spatial and visual dynamics are generated by a free plane that travels through the space and by color, which reinforces the perception of different volumes. The waving surface acts, depending on the context, as ceiling, wall or floor of offices and meeting rooms, guaranteeing visual continuity, movement and flow.

Another important asset to the project is the introduction of several small social and meeting spaces, named silent rooms, which allow for personal retreat, as well as informal meetings or resting. These spaces are intended to generate a highly creative environment promoting comfort and well being among the researchers.

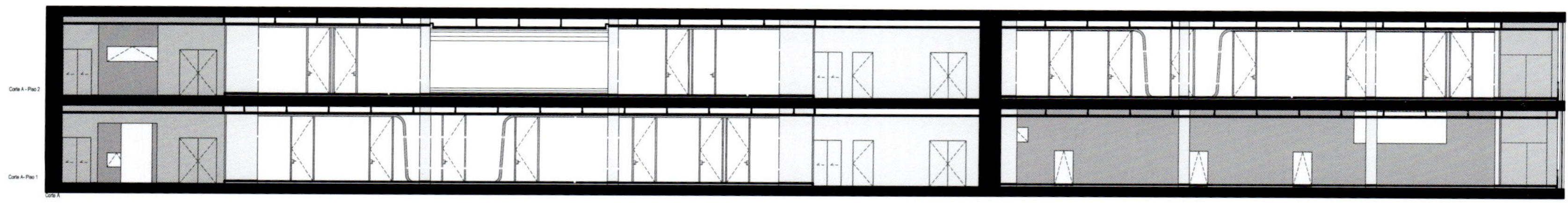

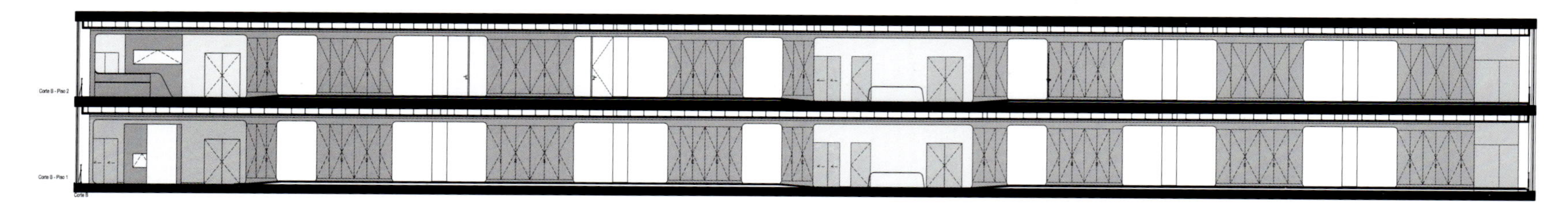

sections

葡萄牙弗劳恩霍夫协会是一个非盈利性的私人研究协会，也是欧洲最大的应用科学研究机构——德国弗劳恩霍夫应用研究促进协会的一部分。尽管不为大众所熟知，但是弗劳恩霍夫协会取得了许多重要的创新成果，包括MP3的文件格式和工作场所组织研究的众多先进技术。

通过公开竞争，Pedra Silva Architects建筑事务所脱颖而出，负责设计其位于波尔图的新总部。新总部位于科技大学园区——波尔图科学与技术校区（UPTEC）。Pedra Silva Architects的设计充分参考了弗劳恩霍夫协会的创新哲学，即简单、积极并充满活力的理念。源于弗劳恩霍夫协会在德国斯图加特的创新中心所采用的创造性工作场所布局和结构元素，也是本案设计的重要组成部分，为本案的设计理念增添了新的层面。

新的研究设施总面积1660平方米，占据了UPTEC一幢新楼的两层。流通性是该项目的精髓。所有的空间都沿着玻璃幕墙分布。沿着这条主轴可以到达所有区域。这些功能各异、大小不同的空间皆是通过一个大胆的设计创造并整合而成：用一个波浪状的平面穿过开放的楼层从而创造出不同的空间环境。

这些空间和视觉上的动态感的产生源自穿过空间的自由平面，以及那些加强不同空间认知感的色彩。在不同的环境中，波浪形的表面充当了办公室和会议室的天花板、墙面或地板，保证了视觉上的连续性、运动性以及流动性。

本案设计的另一个特点是设有几个小的社区和会议区，这些叫做静音室的房间可以作为个人休息的场所，也可以用作非正式的会议室或休息室，旨在创造一个具有创造性并能让研究员感到舒适和健康的环境。

1.07
1.06

WOW!
WE ARE RUNNING OUT OF COFFEE

2.08
ates of David Sarnoff
"Work hard, have fun no drama"
BLA

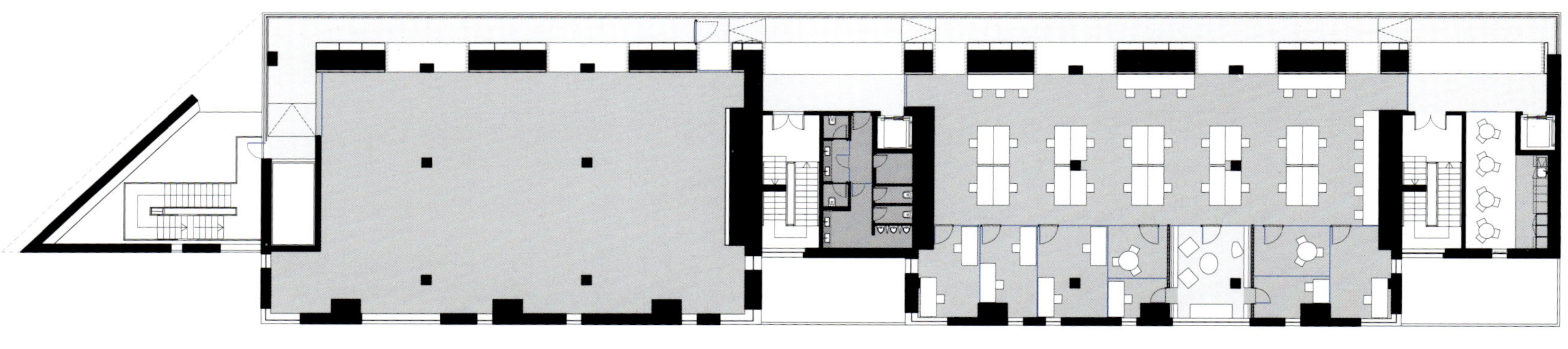

plan 1

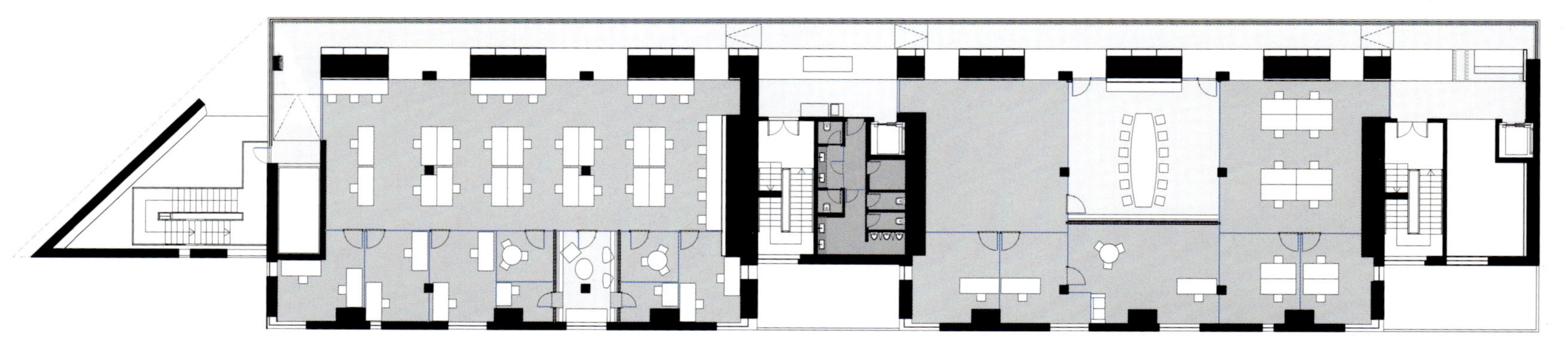

plan 2

"Everything that can be invented has been invented."

Designer Arq. José Lew Kirsch, Arq. Bernardo Lew Kirsch

CORPORATIVO IFAHTO

Firm
ARCO Arquitectura Contemporánea

Associates
Arq. Jonathan Herrejón, Arq. Oscar Sarabia, Arq. Aaron Hernandez, Arq. Yuritza Gonzalez, Arq. Itzel Ortiz, Arq. Nahela Hinojosa, Arq. Anaid Rabanales, Arq. Jose Ruiz, Arq. Federico Teista, Arq. Alejandro Mota, Arq. Andrés Robles, Arq. Julia Villa, Beatriz Canuto, Edna Martinez

Graphic Design
Sociedad Anónima

Rendering
ARCO Arquitectura Contemporânea, Arq. Jonathan Herrejón

Location
Mexico City, Mexico

Area
640m²

Photographer
Jaime Navarro

Team work, balance and flexibility were the elements considered in the design for the new offices of the agency IFAHTO.
The color palette selected for the whole project is plain and neutral colors. The color selection allowed the designers to define very well the spaces, and in the different open areas. Elements such as a green wall, Zen garden, decorative graphics, blackboards and a tilted color wall were incorporated to give the agency its own personality.
The zone marking studies helped define the principal areas: direction offices, operative areas, services, common areas and circulations. The director's offices were located in a strategic way, because each one monitors the activities of his own area. The operative area was done with an open layout to promote team work and make more use of the natural light entrance to reduce energy consumption. The general services were located in the center to have easy access from every point. The dining room was located in the area where the sun light has less effect to maintain a nice room temperature all year long.
The main circulation is on the perimeter and ends in the dining room. The public area black is the predominant color highlighting some white stripes that run from the wall up to the plafond.
The meeting room is a floating box located higher than the rest of the areas. The box is covered with a series of slide walls – that look like scales – covered with graphic images that provide movement and privacy at the same time. Two functions are covered while going into this area: high impact on the visitors– because of the use of the scales – as well as a waiting hall. Going up the ramp and inside the room a big open glass façade welcomes with great ample and fantastic views of the interior space.
Only 3 kinds of finishes were selected for the floors: laminate for the direction offices, carpet for the meeting room and epoxy coatings for the rest of the areas. There is a limited use of plafonds to make more use of the height and the slabs were left apparent to emphasize the ambiance.

ifahto

what if
we take care

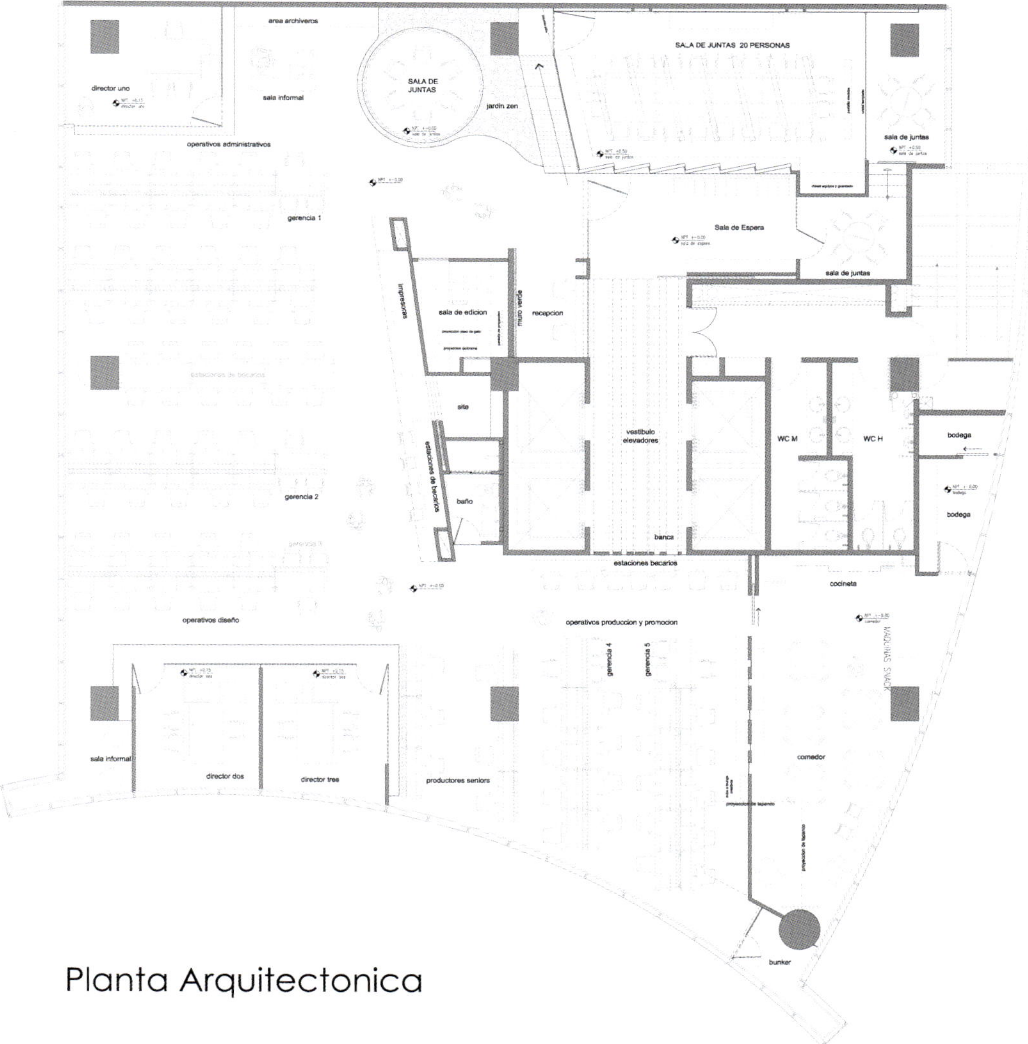

Planta Arquitectonica

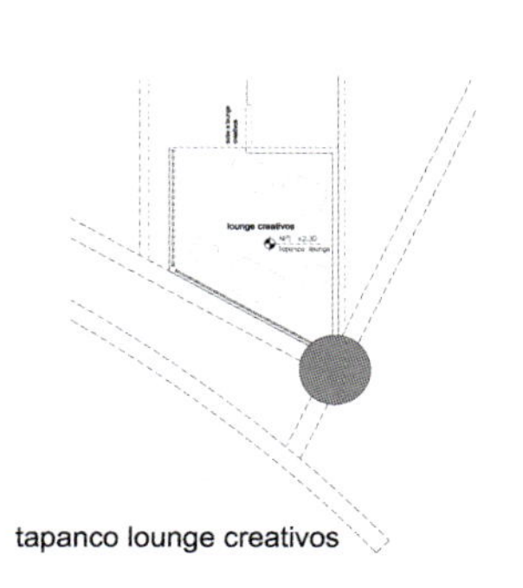

tapanco lounge creativos

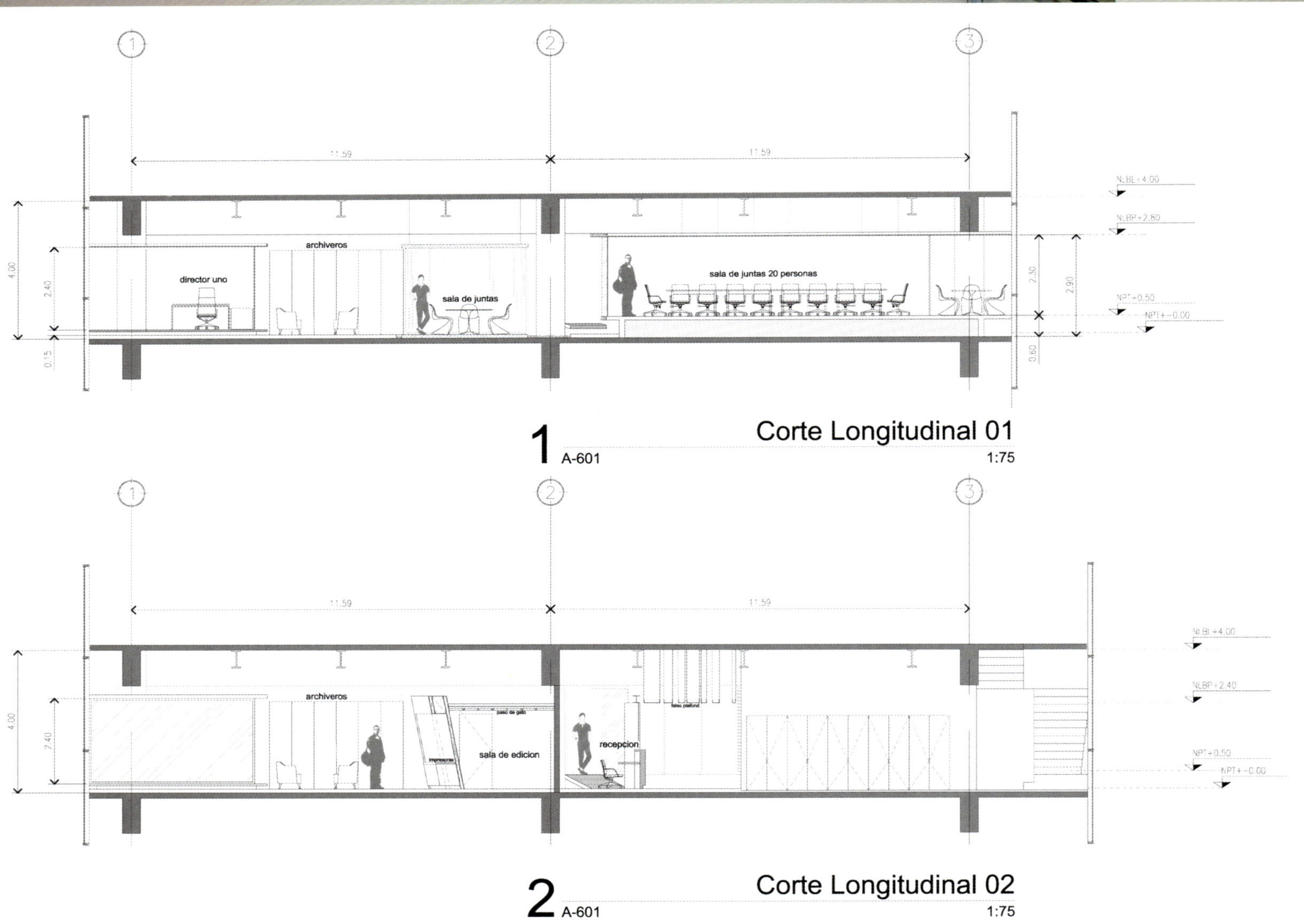

director uno
archiveros
sala de juntas
sala de juntas 20 personas
11.59
4.00
2.40
0.15
2.30
2.90
0.60
N.BL+4.00
N.BP+2.80
NPT+0.50
NPT+-0.00
1 A-601
Corte Longitudinal 01
1:75
archiveros
sala de edicion
recepcion
N.BL+4.00
N.BP+2.40
NPT+0.50
NPT+-0.00
2 A-601
Corte Longitudinal 02
1:75

在设计 IFAHTO 公司的新办事处时，设计师充分考虑了团队合作、平衡性和灵活性这些因素。

整个项目的色彩搭配选择了素色和中性色。这样的选择使设计师能够充分定义空间和不同的开放区域，例如：一面绿色的墙、禅宗花园、装饰的图片、黑板和一面倾斜的色彩墙，这些元素共同赋予了公司特有的个性。

区域标识研究有助于勾勒出传统区域：主管办公室、操作区、服务区、共享区和流通区。因为每个人负责自己区域内的活动，所以主管办公室被布置得非常巧妙。业务区则采用了开放的布局以增强团队合作，并且充分利用射入的自然光线来减少能源消耗。一般服务区位于中心位置以方便为各区域服务。餐厅则位于受阳光照射较少的区域，使全年都能有适宜的室内温度。

主流通区设在周边，一直延伸到餐厅区域。公共区域的黑色主要是为突显从墙面一直延伸到带有装饰的天花板的白色线条。

会议室位于比其他区域都高些的地方，像一个浮动的箱子。箱子被一面面带有圆形图像的滑动墙包住，看起来像鱼鳞，这在提供了动态感的同时也保护了隐私。这片区域有两个功能：通过“鱼鳞”能够带给访客强烈的视觉冲击，同时这里也是等待大厅。从斜坡上去是一间会议室，外部一扇大的玻璃幕墙呈现出宽敞的内部空间和极好的视野。

地板只选择了三种材料：主管办公室采用的是叠层组合材料；会议室选用的是地毯；而其他地方选用的是环氧树脂。局部采用装饰板画以更多地利用空间高度，而使厚的木板露在外面以营造氛围。

what if
we rock

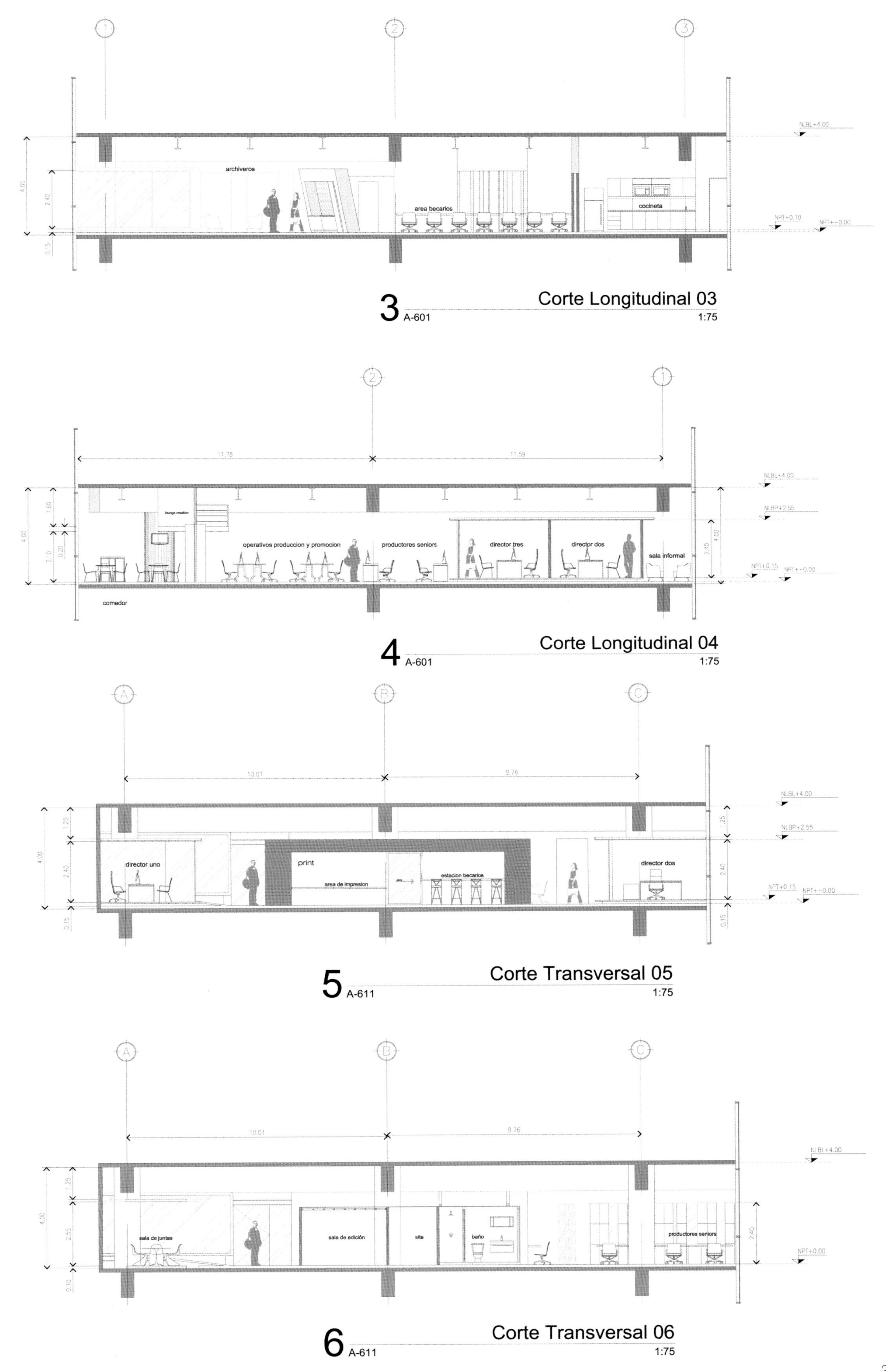
archiveros
area becarios
cocineta
NPT+0.10
NPT+-0.00
3 A-601
Corte Longitudinal 03
1:75
comedor
operativos produccion y promocion
productores seniors
director tres
director dos
sala informal
4 A-601
Corte Longitudinal 04
1:75
director uno
print
area de impresion
estacion becarios
director dos
5 A-611
Corte Transversal 05
1:75
sala de juntas
sala de edición
site
baño
productores seniors
6 A-611
Corte Transversal 06
1:75

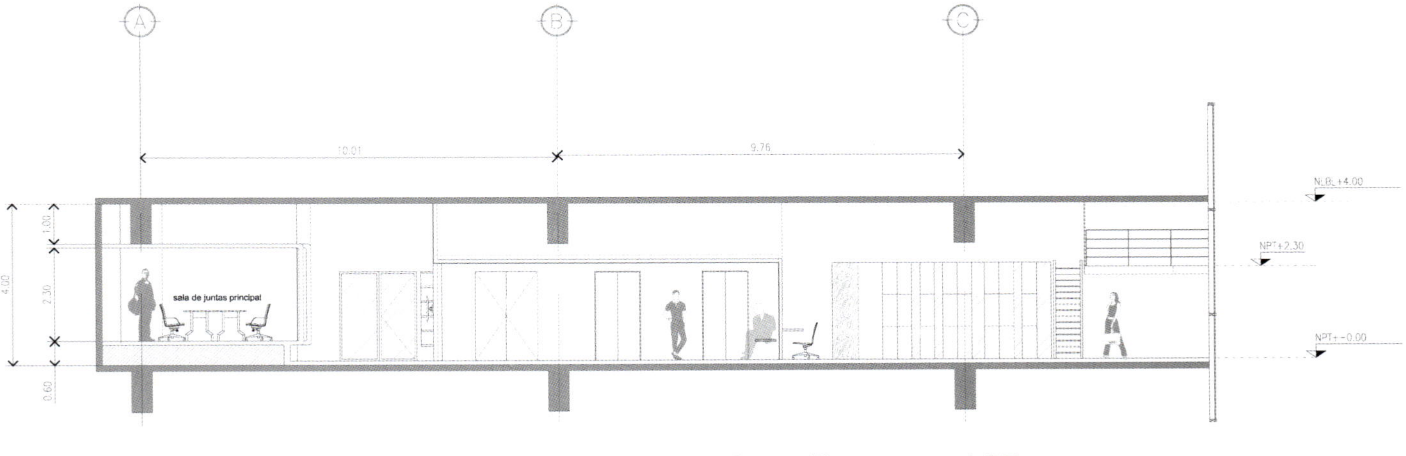

7 A-611 Corte Transversal 07
1:75

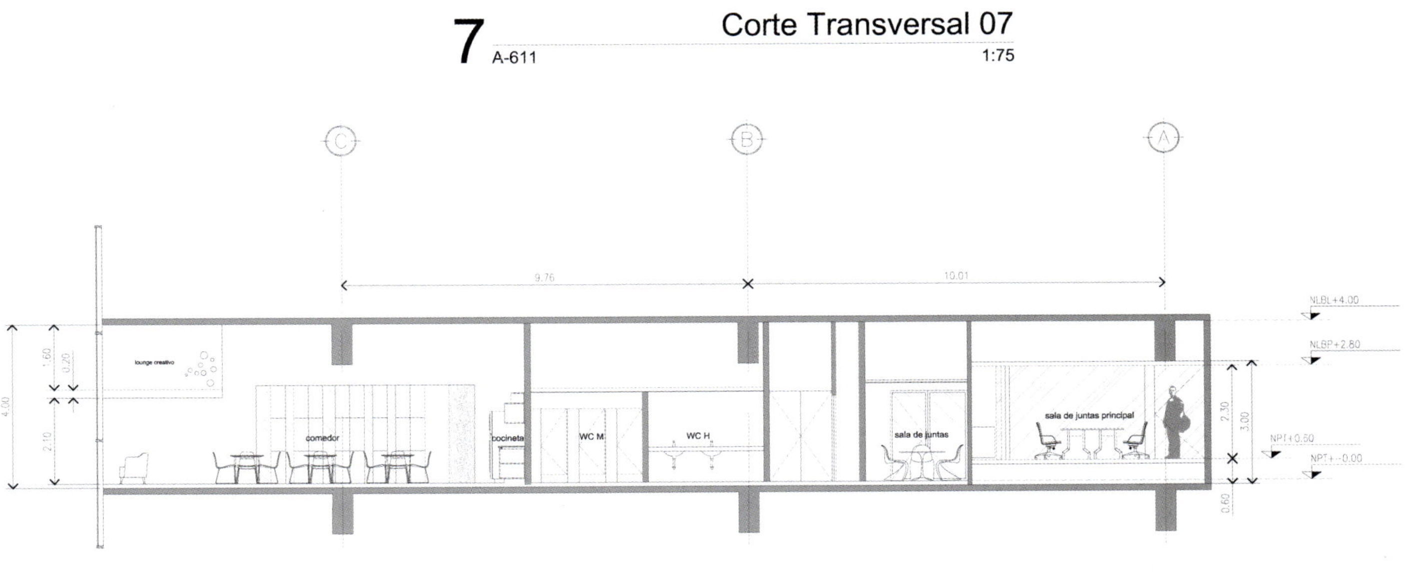

8 A-611 Corte Transversal 08
1:75

Designer stuart martin

SKYPE – CORPORATE HEADQUARTERS

Location
Clausen, Luxembourg

This project, the Corporate Headquarters for Skype, is based in a new building that forms part of the regeneration works to an existing brewery located in Clausen, the old quarter of Luxembourg City.
The challenge set by the client was to provide an interior that would inspire innovation in the daily workplace of Skype employees whilst retaining an element of corporate sophistication and provide high performance AV meeting facilities together with a dynamic flexible office space.
The response was to step back and review current working practices at Skype and develop a wider strategy, which the designers called the Skype DNA.
This would form the blueprint for Skype offices globally. Notably this included delivering a more fluid and dynamic office layout that could support the nomadic, roving nature of many of the Skype employees, whilst encourage a more collaborative approach to their working processes.
With this project the ceiling is employed as a topographical device to encourage movement through the space. The wave like form is a contextual reference to the river running along side that once served the brewery. The ceiling design also fully integrates the lighting and building service systems. Importantly it gives an uplifting ambience to the space. People "look up" when they enter and immediately know they are in a space that challenges office convention with innovative thinking.
The high performance acoustic meeting pods are clad in timber, again a reference to brewing barrels, but also offer a softer and sustainable material to the palette.
The pods are like "Lily Pads" in the river, located to provide a segueing district between office teams encouraging the flow of movement between them.
The building's glazed skin is a ventilated façade system providing efficient building heating and cooling with good quality natural light and views. All materials selected by WAM for the design fit out process have been carefully considered in terms of the environmental impact and sustainability credentials.

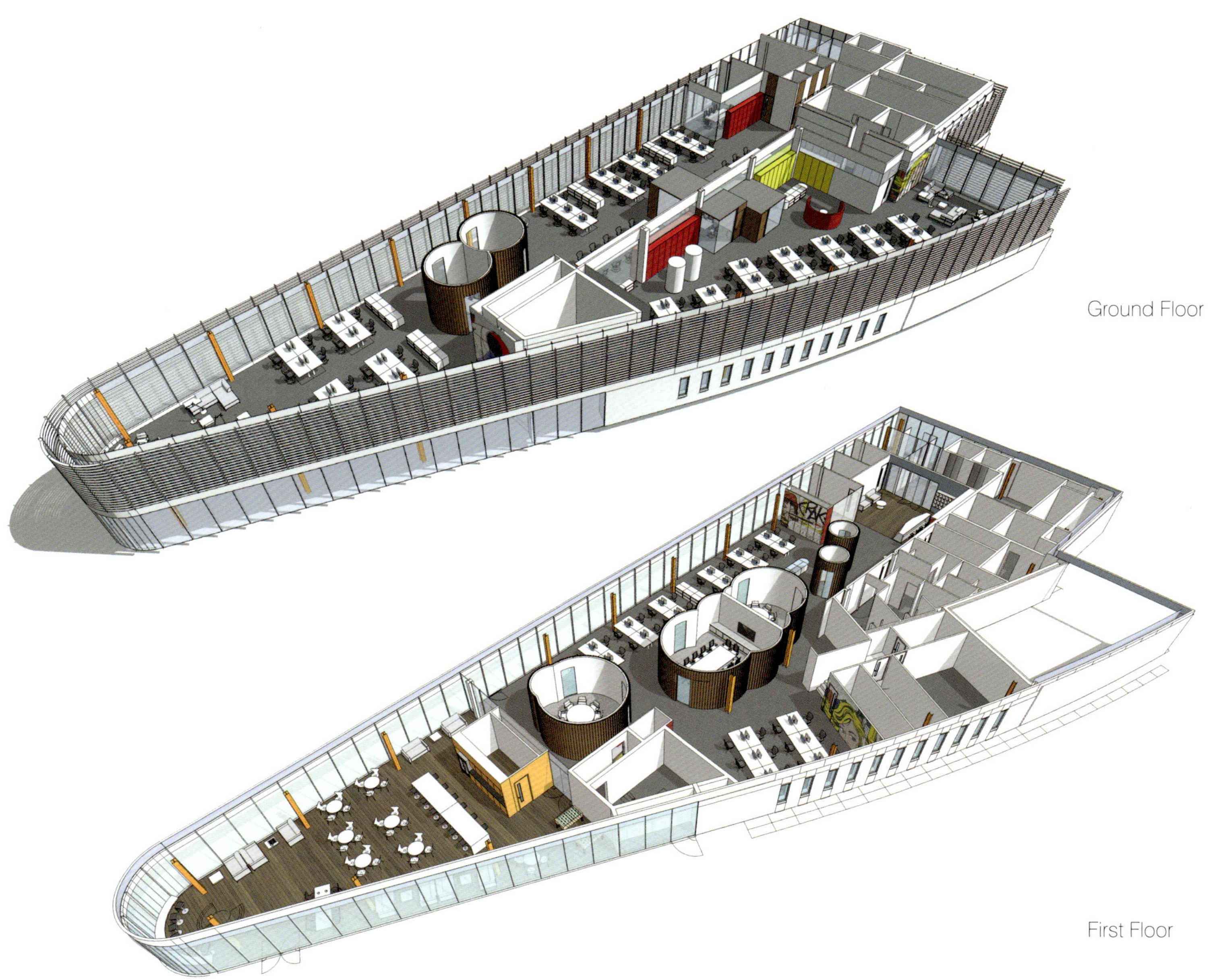

Ground Floor

First Floor

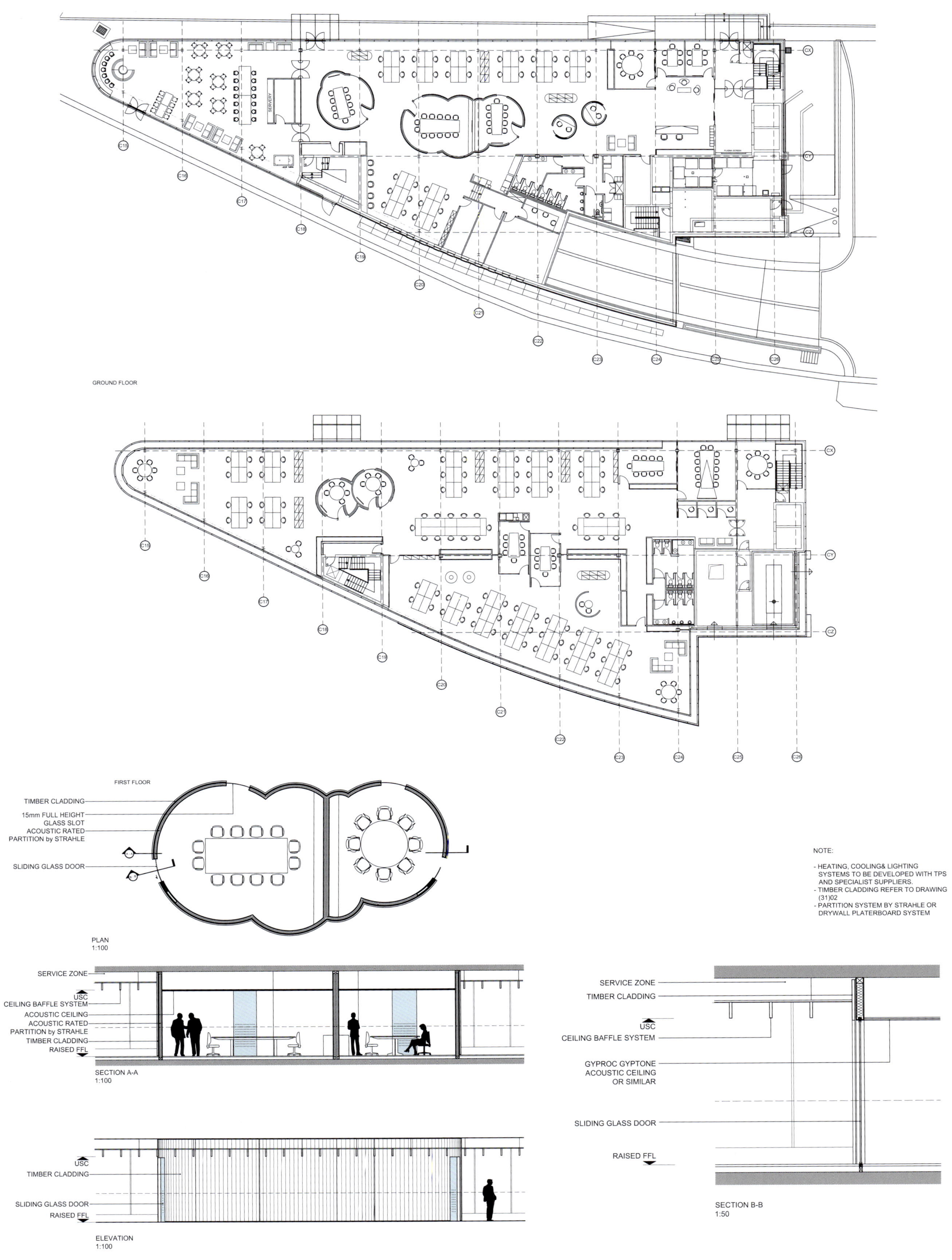

GROUND FLOOR
FIRST FLOOR
TIMBER CLADDING
15mm FULL HEIGHT GLASS SLOT
ACOUSTIC RATED PARTITION by STRAHLE
SLIDING GLASS DOOR
PLAN
1:100
NOTE:
- HEATING, COOLING& LIGHTING SYSTEMS TO BE DEVELOPED WITH TPS AND SPECIALIST SUPPLIERS.
- TIMBER CLADDING REFER TO DRAWING (31)02
- PARTITION SYSTEM BY STRAHLE OR DRYWALL PLATERBOARD SYSTEM
SERVICE ZONE
USC
CEILING BAFFLE SYSTEM
ACOUSTIC CEILING
ACOUSTIC RATED PARTITION by STRAHLE
TIMBER CLADDING
RAISED FFL
SECTION A-A
1:100
SERVICE ZONE
TIMBER CLADDING
USC
CEILING BAFFLE SYSTEM
GYPROC GYPTONE ACOUSTIC CEILING OR SIMILAR
SLIDING GLASS DOOR
RAISED FFL
USC
TIMBER CLADDING
SLIDING GLASS DOOR
RAISED FFL
ELEVATION
1:100
SECTION B-B
1:50

NOTE:
- ALL GRAPHICS TO BE AGREED WITH SKYPE DESIGN.
- CHILLED BEAM&LIGHTING NOT SHOWN FOR CLARITY.
C26
C25
C24
C23
C22
C21
C20
C19
C18
C17
C16
C15
LIFT LOBBY
MEETING ROOM
OPEN PLAN
U/S CEILING BAFFLE
RAISED FFL
ENTRANCE/ LIFT LOBBY
SKYPE BOOTH
MEETING POD
MEETING POD
CHILL-OUT
U/S CEILING BAFFLE
RAISED FFL
BASEMENT CAR PARK
SECTION B-B
SERVICE ZONE
DRYWALL INFILL PANEL
CEILING BAFLE FINS
GLASS OVERPANEL
STORAGE WALL SYSTEM
U/SIDE OF SLAB
USC LEVEL
DOOR HEAD LEVEL
RAISED FFL
3300
2800
2700
STORAGE WALL SYSTEM
MANIFESTATION TO GLASS SEE GRAPHICS/ARTWORK PACKAGE
STORAGE WALL SYSTEM
STORAGE WALL SYSTEM
ELEVATION

NOTE:

- ALL GRAPHICS TO BE APPROVED BY SKYPE DESIGN
- GRAPHICS IMPREGNATED INTO LAMINATE FACIA FINISH
- CHILLED BEAM&LIGHTING NOT SHOWN FOR CLARITY.
- MANIFESTATION TO GLASS SEE GRAPHICS/ARTWORK PACKAGE.

ELEVATION 1
1:50

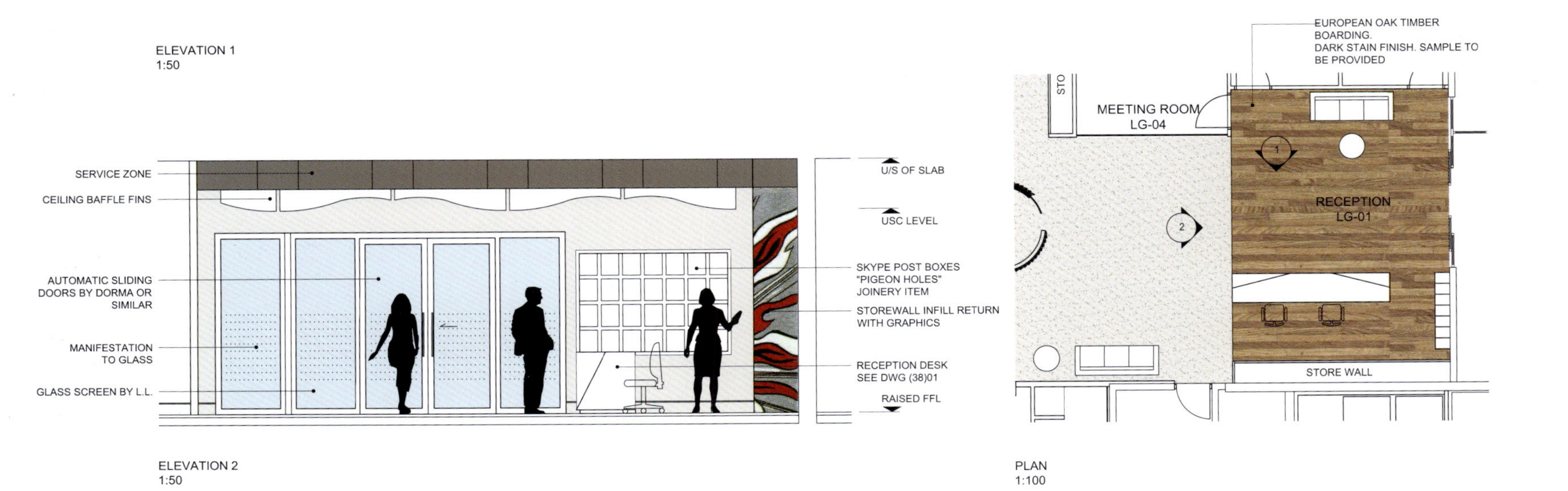

ELEVATION 2
1:50

PLAN
1:100

Johann Gutenberg
Edwin H. Armstrong

Skype 公司总部这个项目位于卢森堡老区克劳森一处现存啤酒厂经部分重建而成的新建筑中。

客户要求设计师在 Skype 员工的日常办公场所创建一个既能够激发创新意识，又能够展现企业文化元素的内部环境，同时还具备高性能音频视频会议设备的灵活办公空间。

鉴于此，设计师将首先熟悉 Skype 当前的工作环境，然后做出一个更宽泛的策略，他们称其为“Skype 基因”。

这个策略为 Skype 全球的办公场所勾勒了蓝图。特别值得一提的是，该策略包括设计出流动灵活的办公布局，既能满足 Skype 员工独立工作的办公环境要求，又能鼓励他们在工作中进行团结协作。

在该项目中，天花板被用作拓扑设备，以增强空间流动性。波浪型的设计灵感源自啤酒厂旁曾为其提供水源的流动河水。天花板的设计也充分结合了灯光及建筑内部的服务系统。关键在于，这样的设计在办公空间中创造了一种令人振奋的氛围。人们一走进办公室，就会立刻发觉自己正身处一个用创新思维挑战传统的办公环境之中。

高性能的听觉会议设备用木料遮盖，其设计灵感源于啤酒厂的酿酒桶，也为平面提供了更软、更持久的材料。

这些设备像是小河里的“睡莲叶”，为鼓励团队之间的互动交流提供了一个共享区域。

建筑的玻璃外墙具有表面通风系统，可高效取暖制冷，也可透入高质量的自然光，人们还可由此看到美丽的窗外之景。所有设计和修建过程中使用的材料都由 WAM 精心挑选，在这一过程中，他们充分考虑了材料对环境的影响和其耐用性。

Designer Shafie Latiff, Shark Rased, James Loh Shixiong

ALLEN & OVERY

Design Company
BBFL

Project Management Team
Cushman Wakefield

Builder Main Contractor
Milan Decoration

M&E Main Contractor
Jag Engineering

Location
OUE Building, Singapore

Area
2,787m²

Photographer
Living Pod

A&O have had an international presence since its foundation in the 1930s and have always seized new opportunities to practice across the world.

Over the years A&O have helped shape and been involved in many ground-breaking developments in the legal field. A&O advised on the first ever hostile takeover in the City of London, drafted the first ever Eurobond in the 1960s and have taken the lead on creating many innovative structures in all core areas of legal practice.

The new office intended for A&O at 50 Collyer Quay has been designed by BBFL based on sound basic principles of developing spaces which besides simply providing the programmatic function the client needs, and also interesting as individual spaces within their own right.

Integration of new technology such as Smart Glass (Auto-frost) allow new dimensions to be brought into the office, where spaces which are open and spacious can become private and intimate immediately at the touch of a button. The addition of acoustically sound systems such as the Jeb glass containment system create spaces which fulfill the A&O's need for visual as well as audio private areas which are multi-function and which also become statements about the company's attention to detail as well as its future proof, forward looking plans.

The design direction was to create a seamless outlook which was in-line with the technology implemented to the office. Clear architectural lines were applied to project clarity in the design. A "floating" reception stone counter was introduced as a fundamental feature to welcome clients' arrival. The reception lobby attached to an observatory deck overlooks the prestigious view of Marina Bay Sands skyline.

Internal office spaces are made open and staff placement has been strategically planned to maintain the visual connection between key and supporting staff.

Throughout, it is clear from the design that there has been a very strong yet subtle attempt to bring in hospitality standards into the common office work environment, raising the bar on what an office environment can be and dispelling myths about what are the acceptable set and tried rules for office and commercial spaces.

ALLEN & OVERY

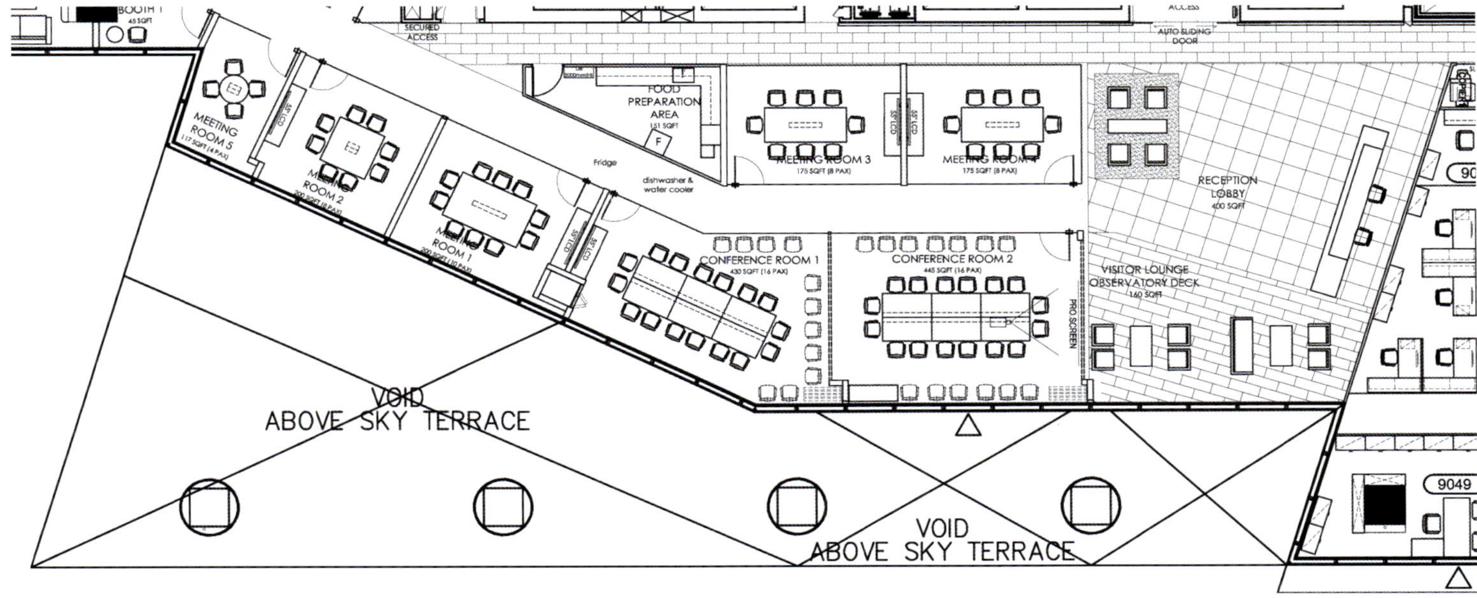

自 20 世纪 30 年代成立以来，时至今日 A&O 律师事务所已经走向了国际市场。但它依然抓紧每一个机遇，以求在世界获得更好的发展。

多年来，A&O 在司法领域取得了很多突破性的进展。A&Q 曾担任伦敦市首次恶意收购的咨询顾问，在 20 世纪 60 年代起草了第一份欧洲债券发行计划，还率先在法律实施的关键领域做了很多创新性建设。

根据发展空间的声音基础原理，BBFL 在哥烈码头 50 号为 A&Q 设计了新的办公场所，不仅满足了客户的工作需求，还在私人空间上发挥了奇思妙想。

诸如智能玻璃（自动冻结）等新技术的综合使用为办公室创造了一个全新的空间。轻轻按下按钮开启智能玻璃，原本开放、宽敞的办公室瞬间就能变成“私人小空间”。声学音响系统，如 Jeb 玻璃密闭系统营造出的空间满足了 A&O 员工对私密的视听场所的需求，不仅实现了办公场所的多功能化，而且体现了公司对细节的关注和远见卓识。

本案的设计方向是打造一个完美的一体化方案——将科技应用于办公室建设。清晰的建造思路在设计过程中得到了体现。体现事务所基本特质的石制“流动”接待台欢迎着客户的到来。接待大厅旁边是一座天文观测台，在那里，你可以鸟瞰著名的新加坡滨海湾金沙。

室内的办公空间呈开放式布局。员工办公场所设计精巧，确保各层次员工能看到彼此。

整体而言，本案的设计明显地透露出想将“友好”的元素融入日常办公环境的微妙意图，提高营造美好办公环境的标准，消除办公和商务空间的固有模式和既定规则。

EXIT

EXIT

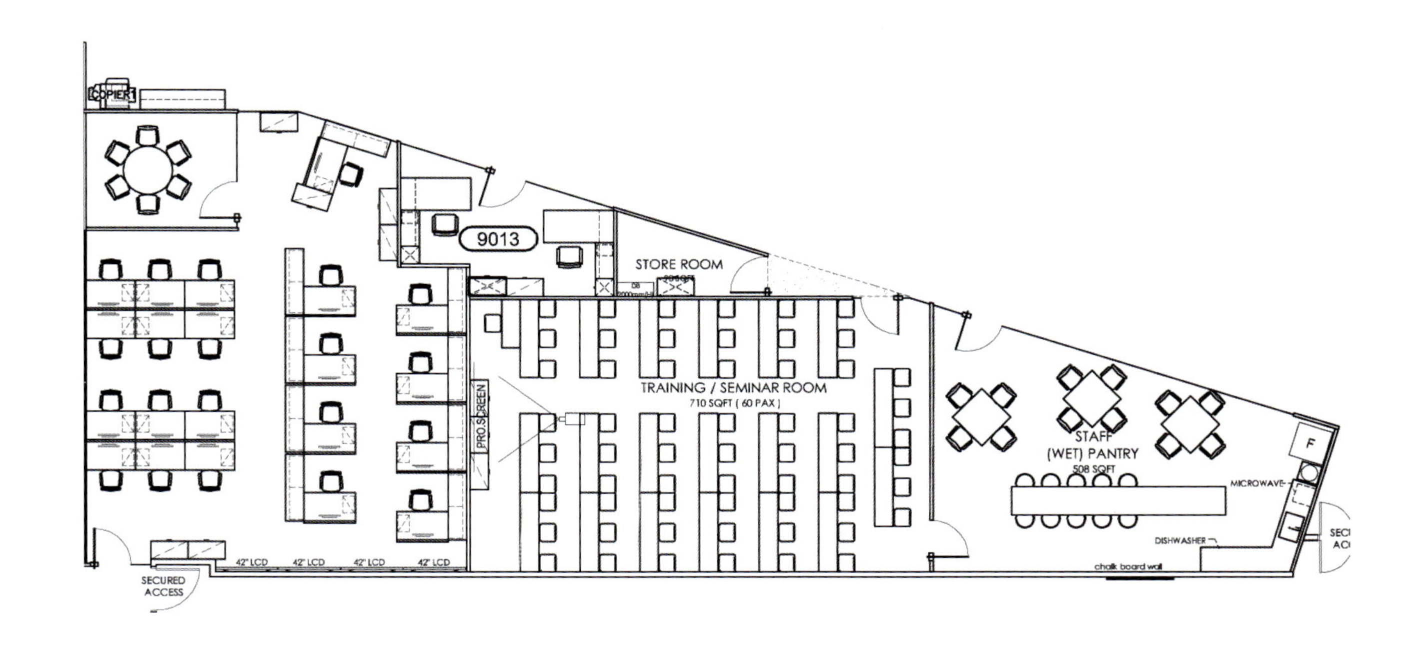
COPIER
9013
STORE ROOM
TRAINING / SEMINAR ROOM
710 SQFT (60 PAX)
PRO.SCREEN
STAFF
(WET) PANTRY
508 SQFT
MICROWAVE
DISHWASHER
chalk board wall
42" LCD
42" LCD
42" LCD
42" LCD
SECURED
ACCESS

Designer Susie Silveri, Anne-Marie Charlebois

ASTRAL MEDIA

Design Company
id+s Design Solutions

Location
Montreal, Canada

Area
1,672m²

Astral is one of Canada's largest media companies. It traces its root in media from Montreal's Greenberg brothers who created a photography chain 50 years ago. Adopting a new brand identity reflecting the company's new corporate profile and evolving culture, along with a move of their headquarters in the heart of downtown Montreal, required the executive floor to be representative of this new vibrant culture and also reflective of the founder's personality and wishes of a warm, comfortable and elegant workspace.

The design challenge was to create a warm, comfortable and "established" executive floor, amidst a highly modern media company identified on the other 5 floors. Reminiscent of finishes of the 1980s, introducing warm light oak wood flooring and paneling in big gestures throughout the space and contrasting it with crisp white walls and dark glass freshly simplifies yet, at the same time, warms the space. Red was used as an accent, not only because of its vibrancy, but also because it was a color that could be both contemporary and timeless. Graphic walls were introduced sparingly to inject color and whimsical humor to the space.

Because of space restraints of accommodating executive window offices and large conference rooms, achieving natural daylight into the reception while still providing visual privacy for the conference rooms was a design challenge. Having no access to the exterior windows and fully-glazed conference rooms not being an acceptable option for this client, created a narrow reception area. The solution to stretch the reception horizontally in reaching out to the windows of the executive work areas in both directions of the building was a key to accessing light and perspectives. This horizontality emphasized by a "waved" feature wall serves as a backdrop to the reception desk and divides the reception area from the main boardroom allowing glimpses to daylight.

During the brainstorming sessions with the employees, they unanimously agreed that the founder is a very jovial, approachable man and a "people-person", but having to be disconnected geographically on the executive floor with his immediate team and conference rooms, and was isolated from the rest of his people. Today, his floor not only reflects his personality and the values and roots of the firm, but allows employees on all the other floors to know that they are warmly welcomed.

astral

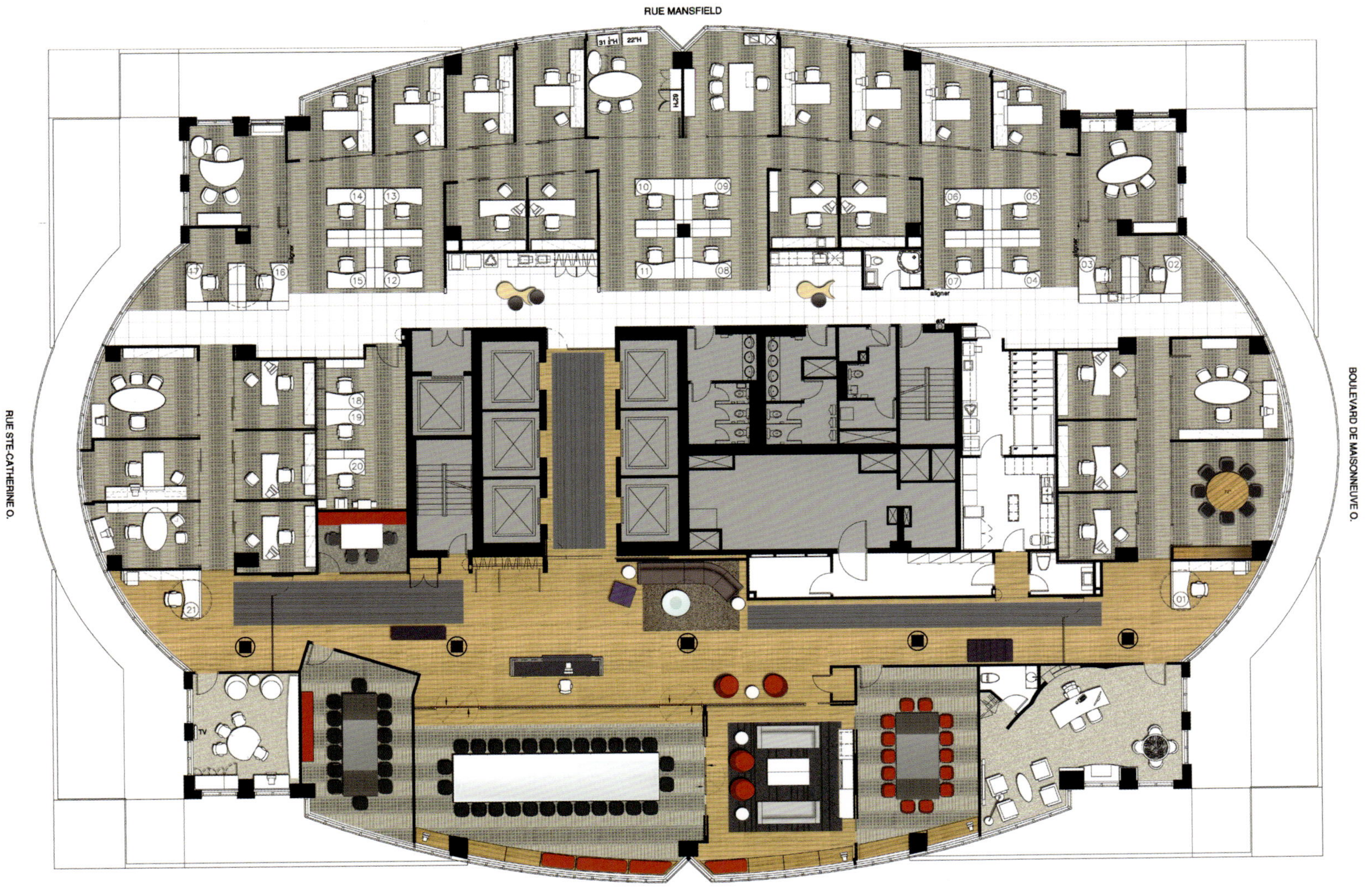

星光传媒是加拿大最大的传媒公司之一。它在传媒业的崛起源于50年前加拿大蒙特利尔市的格林伯格兄弟创作的摄影连锁机构。星光传媒新启用的标识体现了公司的新形象和不断发展的公司文化，同时也纪念了公司总部移址到蒙特利尔市中心的历史。种种新的开始需要星光传媒重新设计其行政楼层，以传达全新的企业文化，彰显创始人的个性，力求打造一个温馨、舒适、雅致的办公环境。

这个设计的挑战是将一座已成型的行政楼层营造出温馨、舒适之感，在一个高度现代化的传媒公司里，让独立的行政楼层在其他5个楼层中脱颖而出。设计师借鉴了20世纪80年代的作品经验，将整个楼层铺上色调柔和的橡木地板和嵌板，搭配明快的白色墙体和黑色玻璃，不仅清新简约，又营造出空间的温馨感。红色做强调色，因为它不仅能散发出勃勃生机，而且具有现代感和持久力。涂鸦墙的使用为办公场所创造了更多的色彩冲击和奇思妙想。

由于楼层面积有限，视窗办公室和大型会议厅的布局安排便受到了限制。既要确保接待室有充足的采光，又要为会议室创造视觉上的隐蔽感，这无疑是设计中的一大挑战。由于无法连接到外部窗口，而且客户也无法接受将会议室全面安装玻璃的做法，整个接待区十分窄小。为了让行政办公区域两边窗户的光线能够照进接待区，水平拓宽接待区域是采光和扩大视野的关键。波浪型墙体作为接待台的背景突出了整个办公区域的水平视觉效果，并将接待区域从主会议室分隔出来以确保阳光的直入。

在和员工们谈话时，设计师得知公司的创始人非常乐观开朗、平易近人，而且很受大家欢迎。但是因为行政楼层布局不合理，他无法及时和团队取得联系，而会议室也将他与其他员工隔开了。现在，他所在的楼层不仅表现了他本人的个性和价值观，也突出了公司的起源，同时也向其他楼层的所有员工张开了欢迎的双臂。

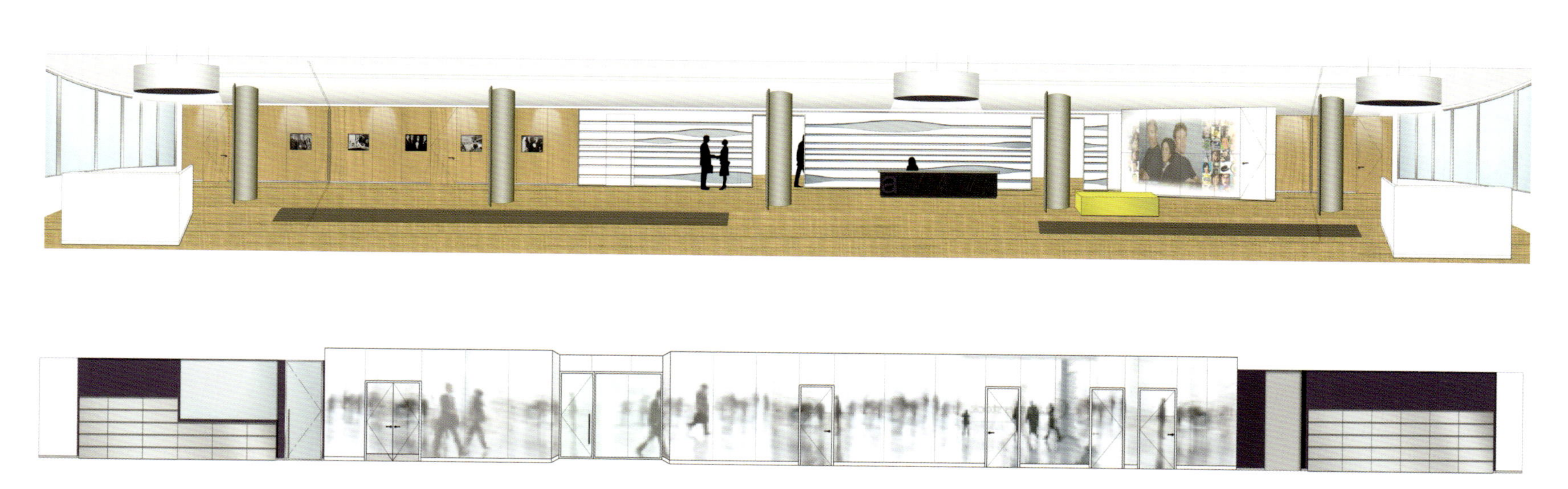

Salle CONECTO

Design Company Robert Majkut Design

OPEN FINANCE

Open Finance is a financial counsel network for individuals, specializing in advising on housing loans and economic counseling. The designers' task was to create a complex image strategy to help make an emerging Open Finance network a credible and recognizable company that would find itself on a strong position among competition. The project's range was wide, and included the creation of corporate design for the network of branches. The branches of Open Finance were to be expressive, modern and characteristic and at the same time they should emphasize dynamic and stable company character. As the symbol of caring about the customers the orchid flower was chosen for the project leitmotif. Orchid embodies protection and attention to unique needs of every single person. It is a metaphor of individual answer to client's expectations. Afterwards, full standardization of network has emerged. The branches were comparable in terms of size and number of customer service points. Modern materials and light coloring of interior were used. White walls contrast with the elegant orange chairs for clients and staff, and with the violet of some furnishing elements. Light curtains made of material with soft, smooth lines give the feeling of privacy. Thanks to eye-catching lightning, emanating optimism and accessibility, branches are easily recognizable. Thus, through design, a completely new value was created – an image of a place corresponding with the character and company values of Open Finance. The project enables foundation of subsequent branches and creates a coherent network's image. It builds Open Finance brand, attracts customers and directly conveys company's modern character. New and recognizable interiors emanate with freshness and openness, in a revolutionary way breaking the convention of a typical financial institution.

open finance
DORADCY FINANSOWI

39
ul. Domaniewska 39
open finance
ul. Domaniewska

openfinance
DORADCY FINANSOWI

STREFA
vip
INWESTOWANIE & KREDYTY

STREFA
vip

STREFA
vip

Open Finance公司是一家从事个人金融咨询服务的连锁企业，专门提供有关住房贷款和理财咨询的服务。设计师的任务是运用多维的形象策略，为Open Finance公司打造出可信度高、识别性强的形象，使其在激烈的竞争中脱颖而出。本案的设计内容广泛，包括系列分支机构的企业设计。Open Finance公司分部的设计既要富有表现力、现代感和自己独特的风格，同时还要突出公司活跃和稳定的个性。兰花象征着对客户的关怀，是整个设计的主题。它体现了公司对客户不同需求的保护和关注，也暗示着公司将根据每位客人的期望来提供建议。而后，连锁公司的完备标准出现了，它们在规模和服务点的数量上基本类似。设计师采用了现代材料和浅色调的设计：白色的墙面和为员工和客户准备的高雅橙色座椅形成了鲜明对比，还有一些紫色系的陈设元素。以带有柔和平滑线条的材质制成的浅色窗帘给人一种隐私感。抢眼的照明设计散发出乐观与亲切的感觉，使公司特征明晰。就这样，通过精心的设计，一个彰显Open Finance公司特点和价值的办公室建成了。此次设计为之后的分支机构的室内设计奠定了基础，也创造出连贯、统一的公司形象。本案树立了Open Finance公司的品牌，吸引了顾客，也直观地表现出公司的现代特点。崭新而富有特色的室内设计散发出新鲜与开阔感，革命性地打破了传统的金融公司的设计。

Designer David Gianotten(Partner in charge), Karbi Chan, Yin Ho, Katja Lam, Mike Lim, Ted Lin, Catherine Ng, Jesung Park, Elaine Tsui, Patrizia Zobernig

MCKINSEY & COMPANY HONG KONG OFFICE

Design Company
OMA

Project Architect
Alejandra Blanco Ackerman

Main Contractor
EDM Construction Ltd.

Acoustic Consultant
Shen Milsom & Wilke Ltd.

Furniture
EDM Construction Ltd., USM, Herman Miller

MEP Consultant
Ferrier Chan & Partners

Electrical Works
Cheung Hing E&M Ltd.

Plumbing & Drainage, MVAC Works
The Great Eagle Engineering Co., Ltd.

Fire Services
Keysen Engineering Co., Ltd.

Security Consultant
Chubb Hong Kong Ltd.

AV Consultant
Ultra Active Technology Ltd.

Location
40/F, ICBC Tower, Citibank Plaza, 3 Garden Road Central, Hong Kong, China

Photographer
Andrew Tang (andrewtang.com.hk), Philippe Ruault, courtesy OMA

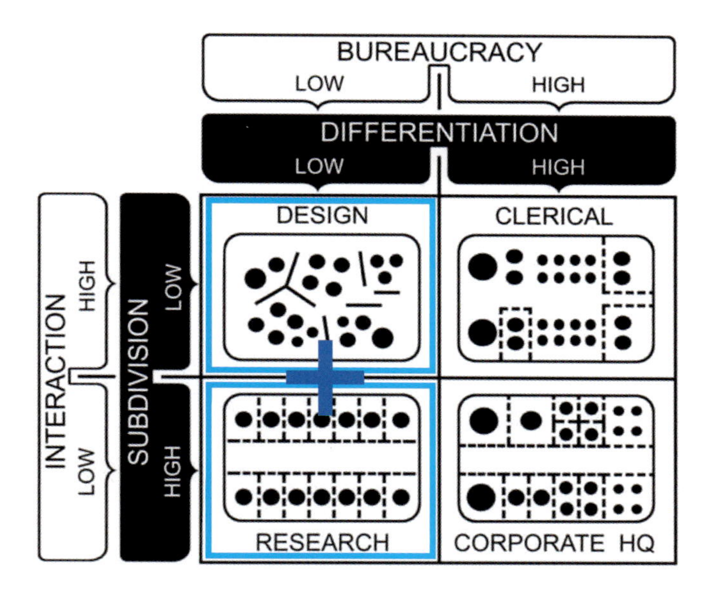

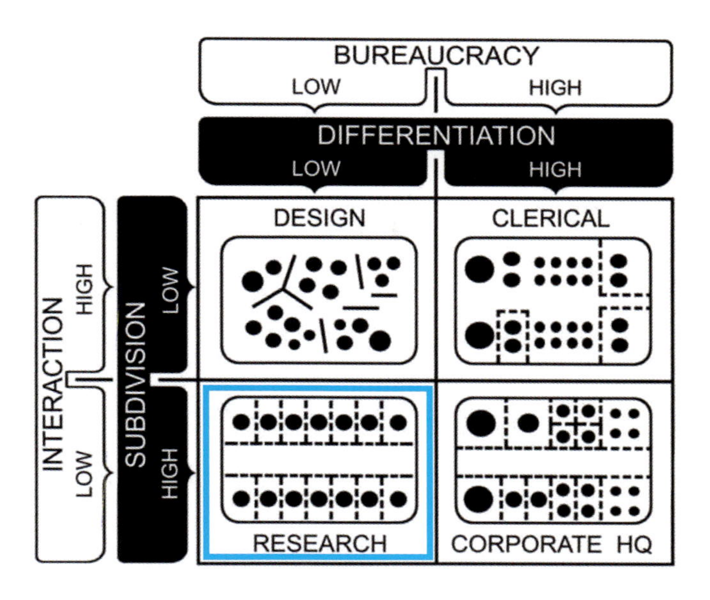

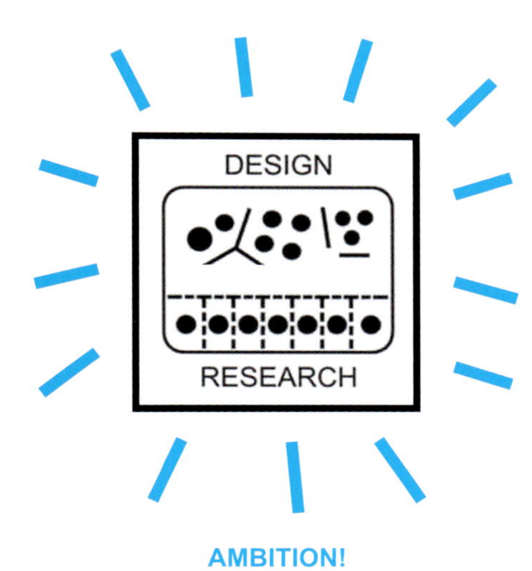

AMBITION!

The new office plan draws inspiration from the black bands on a universal barcode. Different functional spaces are organized in a set of horizontal bands arranged across the office. This design deviates from the traditional corporate office by emphasizing openness yet allowing for confidentiality where needed. Each band respectively accommodates rooms for partners, research teams, staff and clients. Rooms are no longer isolated cubicles solely occupied by one person, but rather a space that different staff members can share depending on their needs. Double glazed glass walls enhance the openness of the office while providing the levels of privacy that the client needs.

The band above the curved bay overlooking prosperous Central employs an open plan for both travelling consultants and some of the full time Hong Kong staff. The travelling consultant can choose where to sit when in town; while Hong Kong based staff have permanent seating. Flexible seatings encourages efficient utilization of office space while accommodating the needs of a highly mobile consulting staff. The openness of the area also encourages interaction among the staff, echoing the spirit of teamwork that is central to how McKinsey works internally, as well as with clients.

The central band, with common areas for staff of all levels, is dedicated to promoting interaction between all staff members and cultivating a stronger sense of belonging. The warm wood of the reception area, fashioned into a playful tree pattern, welcomes visitors as they step into the office. This tree pattern extends into the office, suffusing the main work area with a strong hint of nature. The lounge at the eastern end of the central band, boasting a stunning view of Victoria Harbour, offers the best location for McKinsey Home Fridays. This monthly event gathers the frequent travelling and the Hong Kong based staff to share their experience as a unified office. During regular work days, the staff can bring their laptops to the lounge and work while enjoying the Harbour view. At the other end of the central band is the Quiet Area, a secluded corner for contemplation or rest.

With staff sitting mostly in open areas, a feature of the new design is the addition of several dedicated spots for private conversations. Four circular glass telephone booths are located throughout the open area and lounge for this purpose. The phone booths glow red or orange depending on their vacancy. The colors not only add life to the neutral palette of the office, but also serve the functional purpose of letting staff know when a booth is available. A larger phone booth is provided for conference calls requiring more space for participants.

The new McKinsey & Company Hong Kong Office accommodates the needs for both privacy and interaction, promoting efficiency in terms of the use of space while boosting staff productivity as well as their sense of community.

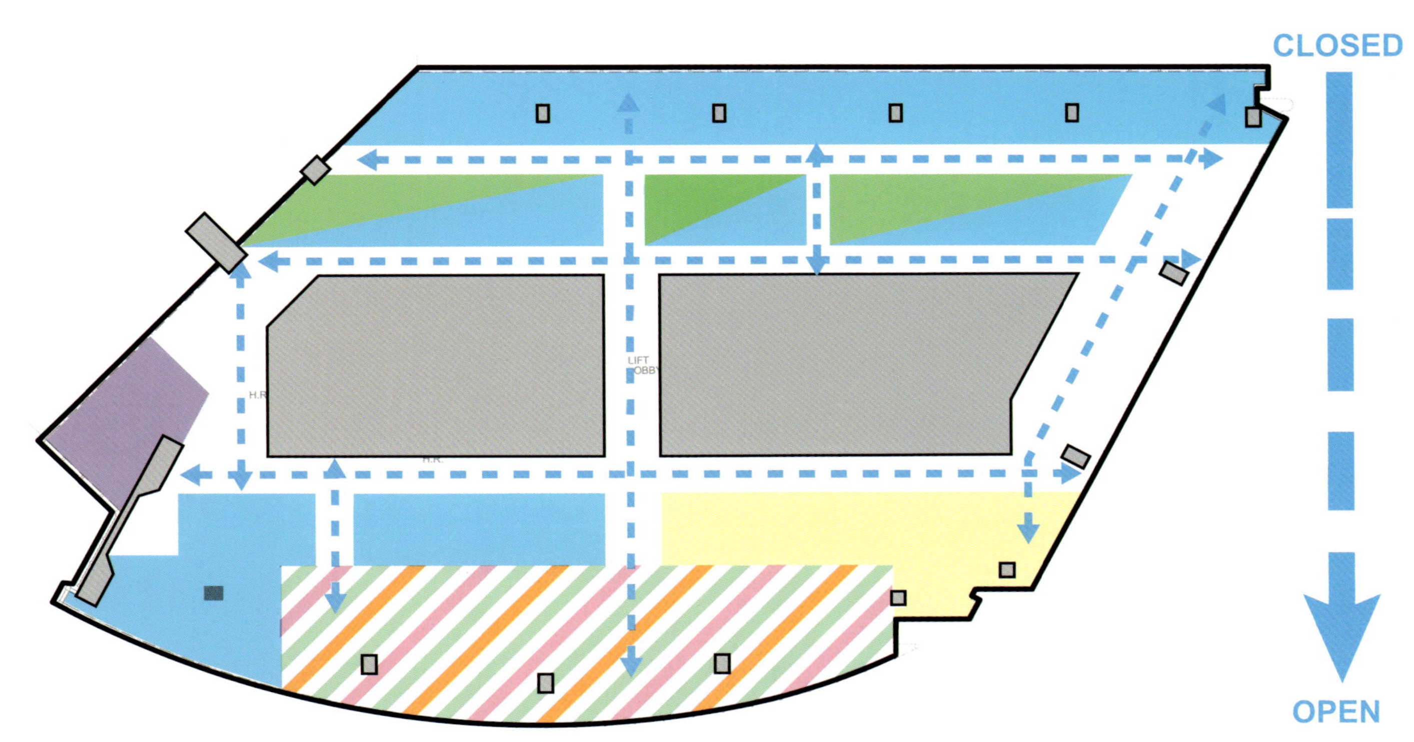

lenovo
Mirra

新办公室的设计从普通条形码的黑色条纹中获得设计灵感。不同的功能空间呈水平带状分布，贯穿整个办公室。这样的设计摆脱了传统的写字楼模式，强调开放空间的同时又在必要时保证了空间的私密性。每个带状空间都为合作伙伴、调研小组、员工和客户分别设置了房间。这些房间不再是单人的小隔间，而是员工们根据各自需要，可共同分享的空间。墙面由双面上釉玻璃制成，增强了办公室的开阔感的同时，也满足了客户对私人空间的不同需要。

在弧形凸窗上面的带状空间，人们可以俯瞰繁荣的城市中心景观，并为来此出差的顾问和一些本地工作的员工提供了敞开式空间。出差的顾问可以随意选择座位，而香港本土员工则有固定位置。灵活的座位有助于有效利用办公空间，也满足了经常出差的顾问们的需要。开阔的空间也有利于员工之间的交流，这与麦肯锡内部的团队精神和与客户沟通的作风相一致。

中央区域的带状空间是所有员工的公用空间，既可以促进员工之间的交流又可以增强员工的归属感。接待区的木质墙面给人一种温馨感，墙面的图案设计成有趣的树状，迎接访客的到来。树状图案一直延伸至办公区域，使其充满了自然气息。在中央工作区东端的休息室，可以欣赏到维多利亚港的美景，同时这里也为麦肯锡 Home Fridays 的举办提供了最佳场所。这个每月一次的活动聚集了经常出差和香港本土的员工，一起分享他们的经验。在一般的工作日，员工们可以带着他们的笔记本电脑去休息室，一边工作一边欣赏维多利亚港的美景。中心工作区的另一端是“无声区”——一个可以用来冥想和休息的隐蔽角落。

由于员工大部分时间都在开放的空间办公，新办公室里特别为员工设计了可以进行私人谈话的专用地点。安装在开放空间和休息室里的四个圆形玻璃电话亭就是为此而设。电话亭会根据使用状态而亮起红色或橙色的灯。这些亮丽的色彩不仅为中性色调的办公室增添了生气，也可以让员工清楚地知道电话亭的使用状态。大一些的电话亭是为多人进行电话会议而设计的。

麦肯锡香港分公司办公室的设计不但满足了客户对隐私和交流的要求，而且能使员工充分利用空间从而提高工作效率，增强归属感。

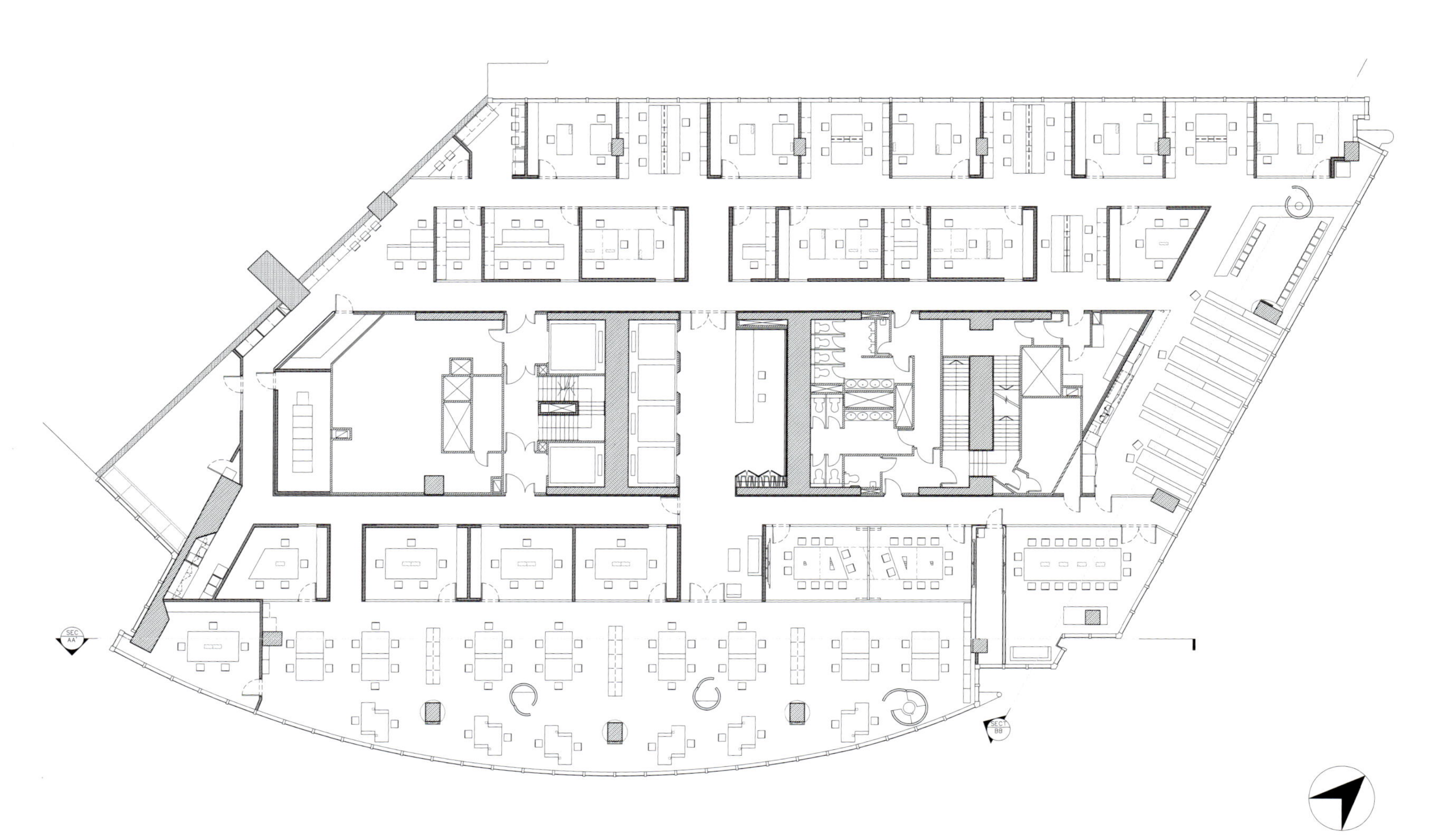
SEC AA
SEC BB
SCALE: 1 / 200
0 1M 2M 4M 8M

Custom Furniture
Stock Furniture

BUSINESS

FORTUNE
FINANCIAL
BUSINESS

Designer Gabriel Salazar, Fernando Castañón

PAGA TODO

Design Company
usoarquitectura

Location
Mexico City, Mexico

Area
2,000m^2

Main Material
Drywall, glass, millwork and carpet tiles

Photographer
Héctor Armando Herrera

Space is the main factor that determines the interior design and operation of a company. The new corporate office for Paga Todo presented a particular challenge because it was necessary to adapt to the clients' demands and a 2,000 sqm area in a shopping center.

A big wood box, inserted respecting the surrounding design, greets everyday collaborators and visitors. Inside the box were located the reception, support area and interview halls, on top of it – with a panoramic view of the finance area – the personalized area to serve the dealers.

The client decided to implement in their office a lounge style cafeteria – like a hotel lobby – because before the relocation the majority of the collaborators preferred to work and meet in the nearby cafeterias to enjoy a more relaxed ambiance. This space has all the necessary services and is a nice surprise for the visitors because there are screen, complimentary computers with Internet, snacks and drinks.

The staircase was located in the vertex of the project in order to communicate with the upper level, opening a new entrance of light from above and making this more interesting meeting point for the colleagues.

The color palette – asked by the client – is very sober and with no risk. White, beige shades with accents in a dry green and the oak of the furniture and woodwork. Three sections with meeting halls divide the space generating references and transitions between the work cells.

For natural light big vertical stripes were open on the façade of the shopping center and most of the walls were not built in ceiling height to make the most of the different natural light sources of the building. Large windows facing the interior of the shopping center were also installed to make references in the main corridor. The windows have random size and create a sequence with the transition of each of the work teams.

空间是决定内部设计和公司经营的主要因素。Paga Todo 公司新办公室的设计带给设计师一个特别的挑战：它不仅要满足顾客的需要，而且要建在购物中心，更需要2000 平方米的巨大空间。

本案设计了一个巨大的木盒子空间，与周边的设计相得益彰，每天接待合作方和来访者。盒子空间的内部是接待处、后勤区和面试厅。盒子空间的上面是专门用来接待经销商的个性化空间，人们在此可以一览金融区域。

客户决定在他们的办公室建一个环境悠闲的自助餐厅，就像酒店大厅一样，因为在重新装修之前，多数合作者就喜欢在附近的自助餐厅会面工作，享受其中轻松的氛围。这里提供所有必要的服务，并且还有大屏幕、免费的电脑和网络、零食和饮料，这将使来访者感到非常惊喜。

为了方便与上层工作人员沟通，楼梯被安置在空间的顶端。这样的设计不仅可以使光线从楼梯口照射下来，也为员工们创造了一个有趣的会面地点。

客户要求的色彩搭配朴实无华。设计师为家具和木制品涂上白色、带有米黄色阴影的干绿色和橡木色。这样，带有会议室的三个区域把空间分隔开来，在不同的工作区形成了标识和过渡。

为了引入自然光线，在购物中心的正面设置了许多巨大的垂直线条，大部分墙面没有建到天花板那么高，以使建筑充分利用自然光源。面对着购物中心内部的窗在主廊中起到参考标识的作用。窗的大小不一，在每个工作团队经过时可出现一组影像片段。

Design Company Hofman Dujardin Architects

DLA PIPER

Location
Amsterdam, the Netherlands

Area
10, 500m^2

Photographer
Matthijs van Roon

DLA Piper is one of the world's largest law firm, with an Amsterdam based office, consisting of 130 fee earners and a total 250 employees. The office is divided to front and back offices. The entrance, coffee bar, restaurant, waiting area, boardroom and meeting center are located in the front office on ground level, mezzanine level and first level. The back office is on the second and third level, with its working rooms, sharing spaces, bar and library.

The front office is located on ground level, mezzanine level and first level. The first space one experiences when entering the DLA Piper office is the two-storey high entrance. Here, the welcome desk is moved from its previous position to the entrance zone. This object is a strong, basic shape made of Corian and is emphasized by the floor pattern. The monumental glass stairs going up in this space are highlighted by the expressive red carpet which lies underneath.

The back office, positioned on the second and third level, has a very strict zoning: the inner ring of working rooms creates the colorful zone, while the corridor and the outer ring establish a natural one. The workers are given a choice for their working environment according to their preferences: colorful modern or classic neutral.

The most excessive spatial changes made to the existing situation can be found at the corners of the building. Not only were all closed rooms removed from here in order to gain more daylight and open up the space for the new working spots. They also create the common areas, the so called "Sharing Spaces". These parts of the building contain facilities such as internal meeting rooms, pantries, a small lounge area and restrooms.

The middle parts of the corridor zones between the corners of the building, have also been opened. This means that the architect didn't place any closed rooms here, but created open areas with new working spots in open space. As a result, the length of the corridor is broken and daylight brings a pleasant atmosphere inside.

Being replaced from the mezzanine level, the office library is located on the 3rd level. Located on the corner of the building, it offers generous space and a charming view to the surroundings.

The new DLA Piper office interior design creates diversity, adventure and tranquility for its employees in thoroughly designed spaces that can stand the test of time.

DLA PIPER

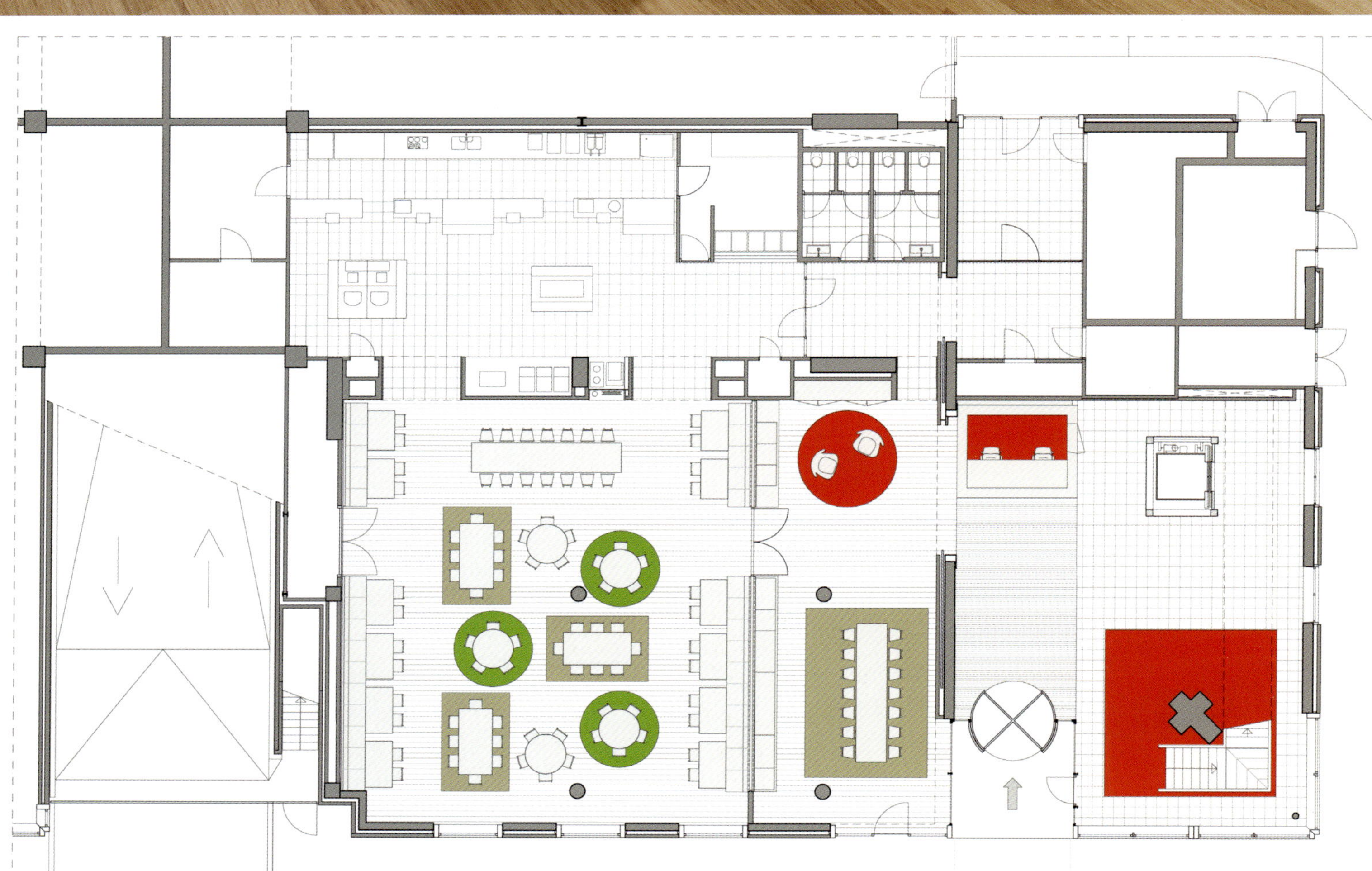

ground floor

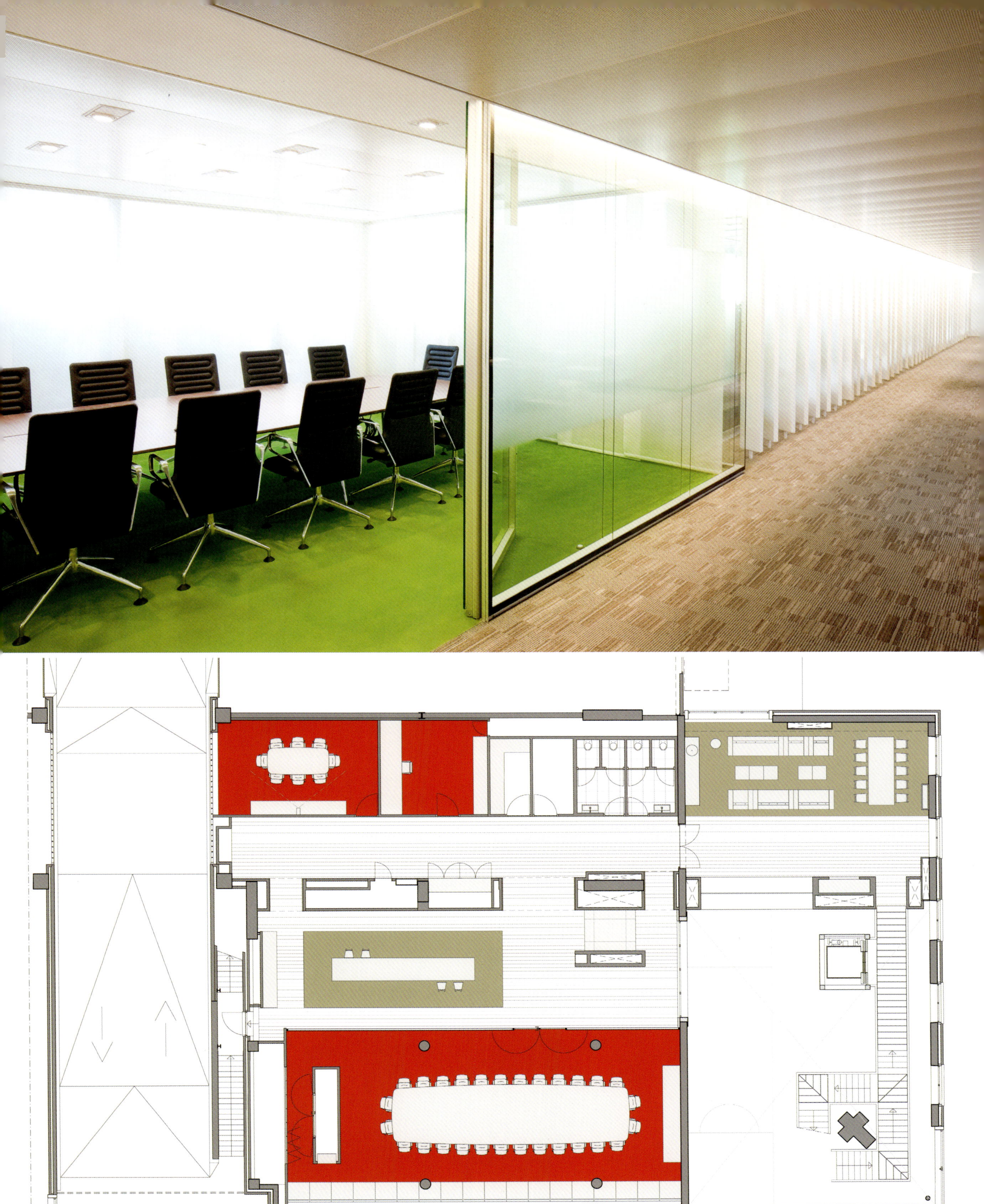

in between floor

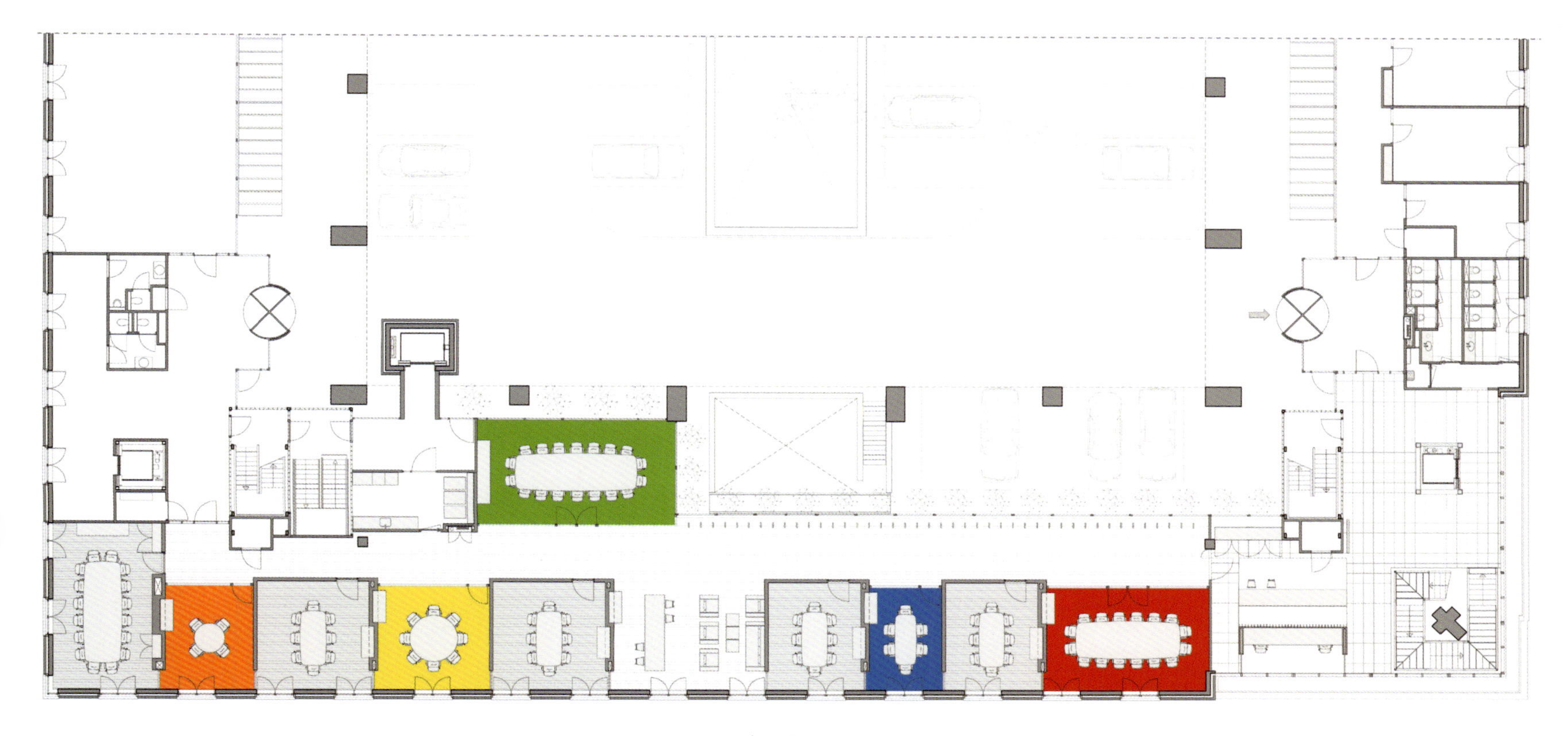

first floor

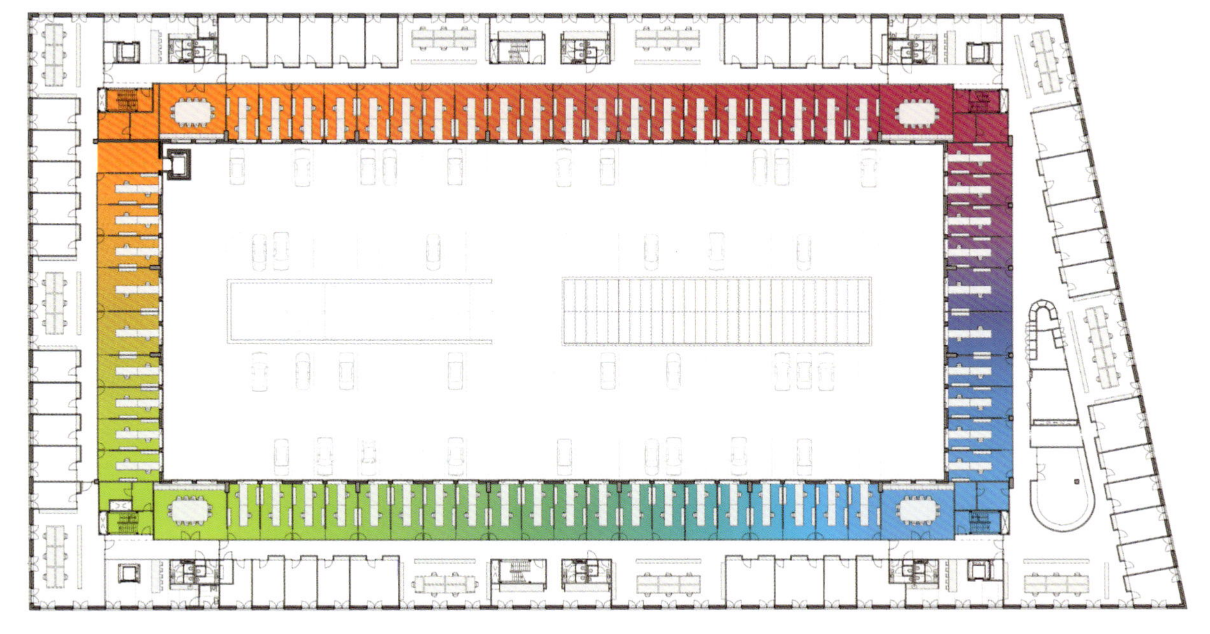

second floor

欧华是全球最大的律师事务所之一，总部设在阿姆斯特丹，拥有130名业务人员和250名员工。总部分为前厅和后厅。入口、咖啡厅、餐厅、候客区、会议室和会议中心都设在一层、夹层和二层的前厅。二层和三层的后厅是办公室、休闲区、酒吧和图书馆。

一层、夹层和二层属于前厅。进入欧华，首先映入眼帘的就是两层楼高的入口。接待台从以前的位置挪到了入口处，这个由可丽耐人造大理石做成的坚硬台面在地面图案的衬托下显得尤为突出。前厅的楼梯由巨型玻璃制成，铺上耀眼的红色地毯，显得格外夺目。

二层和三层属于后厅，布局严谨。内环的工作间色彩艳丽，而走廊和外环则显得相对自然。员工可以根据自己的喜好选择办公环境：色彩鲜艳的现代化风格或是传统的淡雅风格。

空间布局改动最大的是建筑内的边边角角。为了获得更好的采光并拓展空间，设计师不仅将所有的封闭式房间移到了新的办公点，还开辟出一块名为“共享空间”的休闲区，内设内部会议室、食品储藏室、小休息区和卫生间。

大楼内的走廊被设置在拐角处，走廊的中间做了开放式设计。也就是说，建筑师没有在这里设计任何封闭式的房间，而是在这个公共区域建造了新的开放式办公点。这样，走廊在纵向上不再连贯。暖暖的阳光直射进来，整幢大楼都洋溢着愉悦的氛围。

图书馆从以前的夹层搬到了三层的角落里。新的图书馆空间更加宽敞，视角也更加广阔。

欧华律师事务所全新的办公室室内设计为员工打造了一个集多样化、奇特性和静谧感于一体的办公环境，完全能经得起时间的考验。

third floor

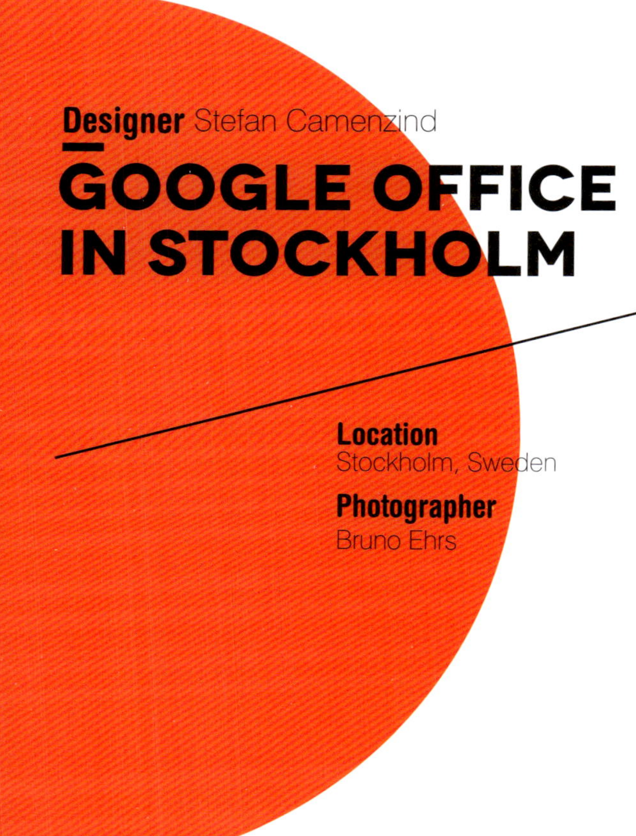

Designer Stefan Camenzind

GOOGLE OFFICE IN STOCKHOLM

Location
Stockholm, Sweden

Photographer
Bruno Ehrs

The new Google office in Stockholm is the successful result of merging three smaller Google offices from different locations across Sweden and Norway into one large combined hub for Engineering and Sales and Marketing departments.
An involvement of the users themselves at an early stage of the design of the project, granted essential insight into the emotional and practical requirements of the Googlers, provided open collaboration, unique perspectives and ideas, and at the same time supplied the local Googlers with a direct sense of ownership of the future office.
Research showed that the local Googlers, despite being part of a constantly growing global company, emphasized the importance of individuality and maintaining a small-company ambience with a strong Swedish identity. As they participated in the design process to define and create their own local identity, the analysis of the users' feedback demonstrated that Sweden with its rural landscapes and the city of Stockholm was of major importance to the definition of their identity.
The diversity of the different backgrounds and personalities required a multi-facetted approach with the creation of different office "landscapes" that would support the Googlers in their work and well-being.
As a result of the research, the Engineering part of the office was created representing the Swedish "countryside" with a more private workspace layout and a comfortable and warm choice of materials and finished. On the other side of the office floor, a colorful and lively open-plan space representing the Stockholm cityscape was developed for the Sales and Marketing teams. Connecting those two different landscapes is the Cafeteria, designed as a "Town Market Square", acting as a meeting point to promote communication and recreation.
On a different lever, the local identity of the office was further accentuated by naming and designing all meeting rooms with different Swedish inventors/ inventions in mind, creating a diverse range of rooms with different designs, sizes and atmospheres, suitable for broad span of meeting requirements.

DELL
A5D2
A5D3

"mf 0.9.3"
major
minor
S

NOBEL
N+
N+
N+

Linné
SYSTEMA NATURÆ
Naturſyſtem

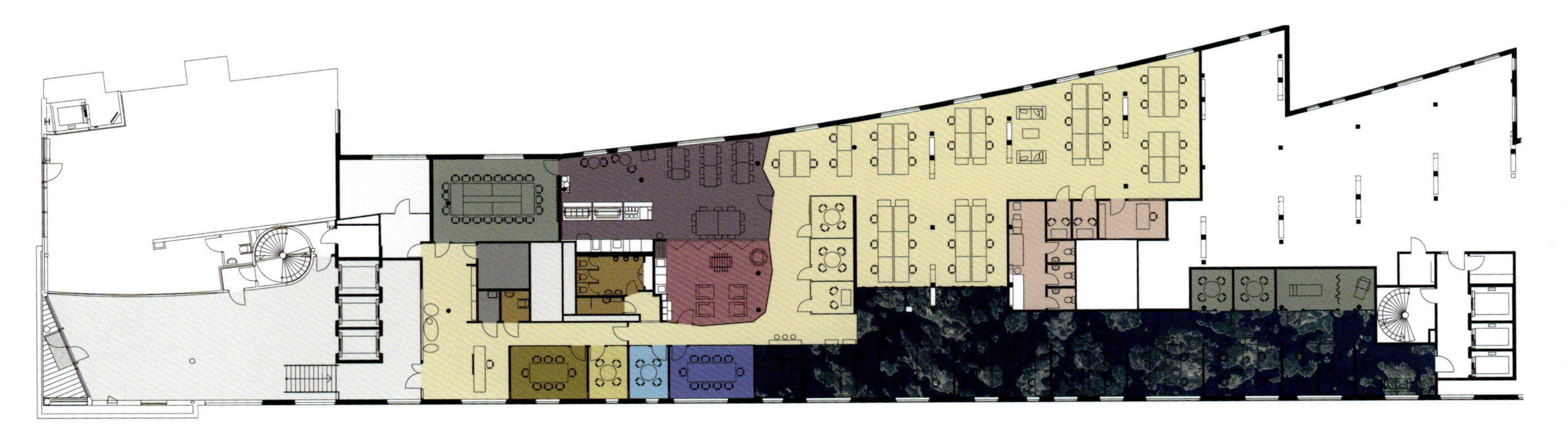

谷歌通过成功并购在瑞典和挪威的三家小办事处，在斯德哥尔摩市成立了新的办公室。该办公室将成为谷歌的工程设计与营销部门。

谷歌的员工可以在项目设计的初期参与到其中，并可以表达自己的情感或实际需求，与设计师坦诚协作，提供独特的视角和思路。同时，这样做也使员工们对新办公室产生一种归属感。

调查显示，尽管作为不断壮大的全球化公司的一员，当地的谷歌员工仍强调个性化的重要性并保持着带有浓重的瑞典特征的小企业氛围。正因为他们参与了界定并创造自身个性的设计，在分析其反馈信息时可以发现，员工对自身个性的定义中最重要的元素就是瑞典的乡村景观和斯德哥尔摩市的重要地位。

由于员工们有着不同的背景以及各自的个性，以多方面的设计手段来创建不同的办公景观，以保证健康、高效的工作环境。

最终的研究结果是：办公工程区域设计了私人化的办公格局，所用材料和装饰以舒适、温馨为主，展现了瑞典的乡村风情；在办公楼的另一侧，设计师为营销团队设计了一个色彩艳丽、生气勃勃的开放空间，展现了斯德哥尔摩靓丽的都市风景。连接两种办公景观的是自助餐厅。餐厅被设计成“镇集市广场”，为员工们提供交流和娱乐的场所。

另一方面，以瑞典发明家或著名发明设计命名的办公室，进一步突出了当地的特色。设计师还设计了一系列不同大小、不同风格和氛围的办公室以满足不同的需求。

Toronto, CA
New York, US
London, UK
Stockholm, SW
Moscow, RS
Beijing, CH
Tel Aviv-Yafo, IS
Hyderabad, IN
Sydney, AS

Pro-Winner
MOST
WELCOME
With lots
Love
GOOGLE ROCKS!
TANYA STEFAN

BELLE
JARDINIERE

Designer Gerd Priebe

SAEGELING

Design Company
Gerd Priebe Architects & Consultants

Location
Heidenau, Germany

Area
600m^2

Based on the existing small scale building structure of the neighborhood, the studio of GERD PRIEBE ARCHITECTS & CONSULTANTS designed an elegantly curved two-storey building structure that despite its deliberate unique form fits harmoniously into the existing city planning context. Aspects of sustainability were considered from the start of planning and played an integral part in the design.

Integral planning, user comfort, visual comfort, acoustical comfort and thermal comfort were incorporated in detail.

The spacious and open character of the new building manifests itself on the east side where the curving glass façade envelopes a secluded outdoor space which visually flows into the indoor space. The glass façade dominates and communicates a heightened sense of openness and light flooded spaces. Two functional areas divide the building. To the south, the reception area or "public space" welcomes visitors and leads them through a two-storey lobby which doubles as event space. Adjoining the area on the ground floor is a meeting room for clients and in the upper storey a gallery is located.

Glass, steel and concrete are the predominate construction materials used to create a homogenous structure. The curvilinear shape of the building optimizes the rigidity and stability of the loadbearing construction and allows for a reduction in materials and wall thicknesses.

The fine, white stucco surface of the façade is juxtaposed with the light green glow of the convex and concave glazing. The curved, laminated safety glass, reaching up to 5.3 m in height, was specially manufactured for the project in Finland. The glazing, the first of its kind in Germany, is sealed vertically without joints and is fixed with a horizontal cover strip. Further distinguishing the unique character of the curved plate glass façade are the individual sections designed with several radii. The exterior areas possess glass panes with up to two radii and in the interior areas there are panes containing up to three different radii.

In configurating an effective office and work area, the strict partitioning of user and circulation space has been disbanded with the objective of creating a flowing, efficient and practically support-free architecture.

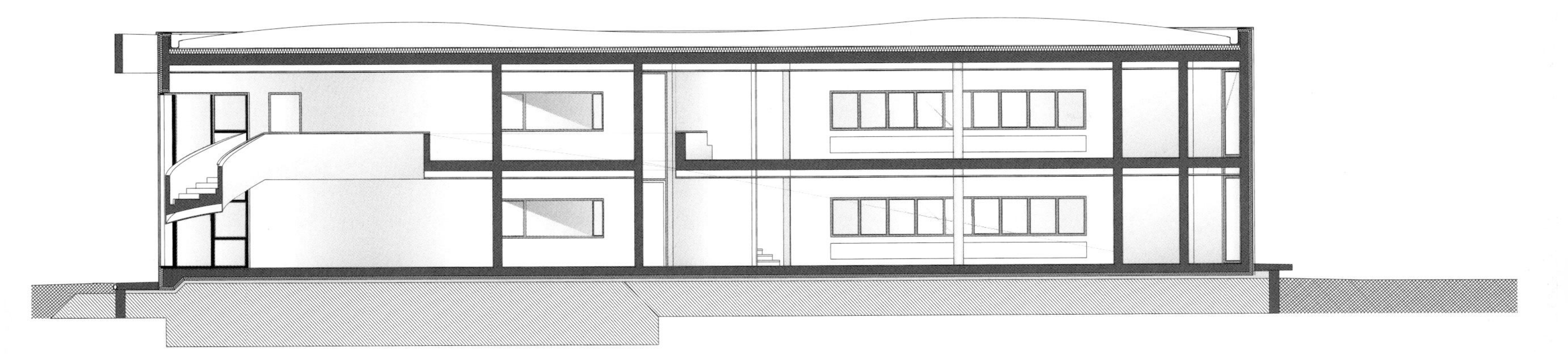

Saegeling Medizintechnik I Heidenau

longitudinal section

Saegeling Medizintechnik I Heidenau

cross section open space office

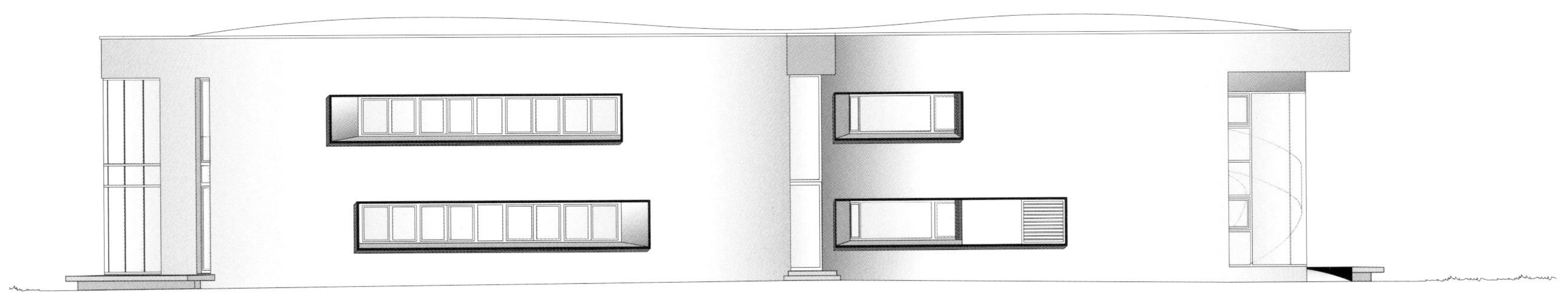

Saegeling Medizintechnik I Heidenau

west elevation – staff entrance

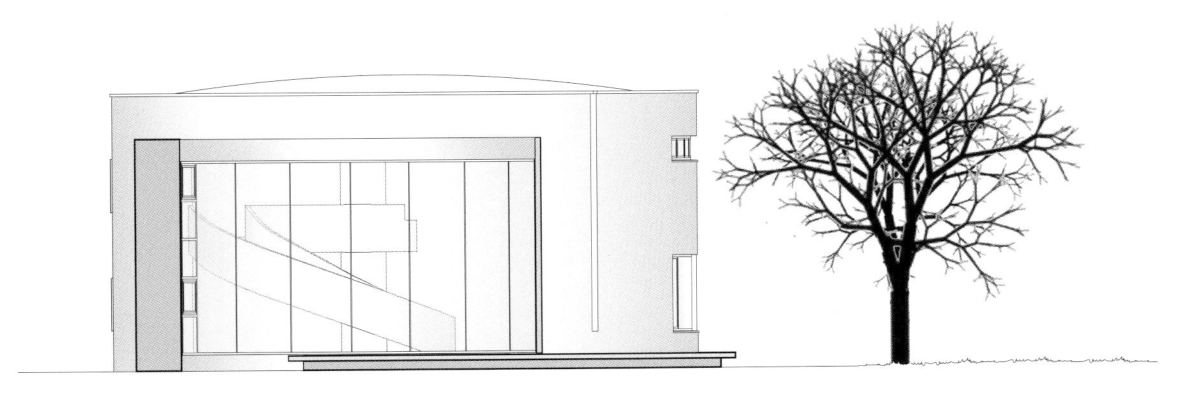

Saegeling Medizintechnik I Heidenau

south elevation – customers' entrance

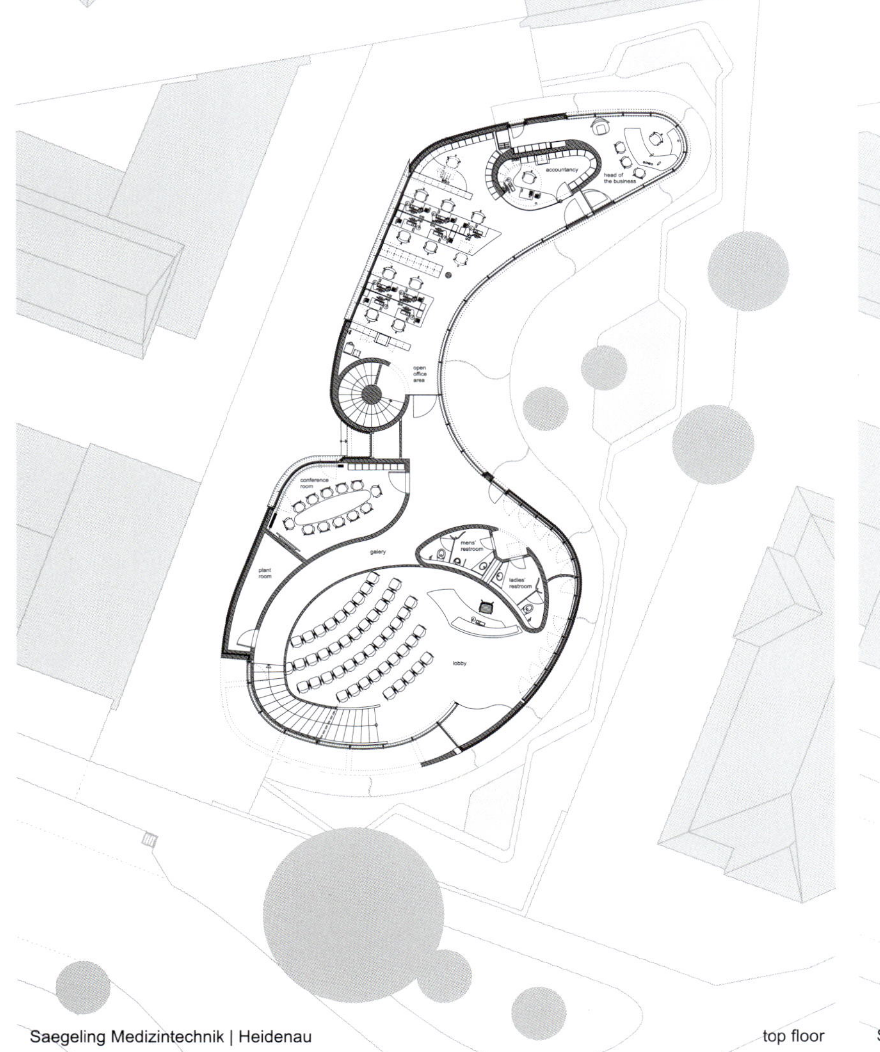

Saegeling Medizintechnik | Heidenau

top floor

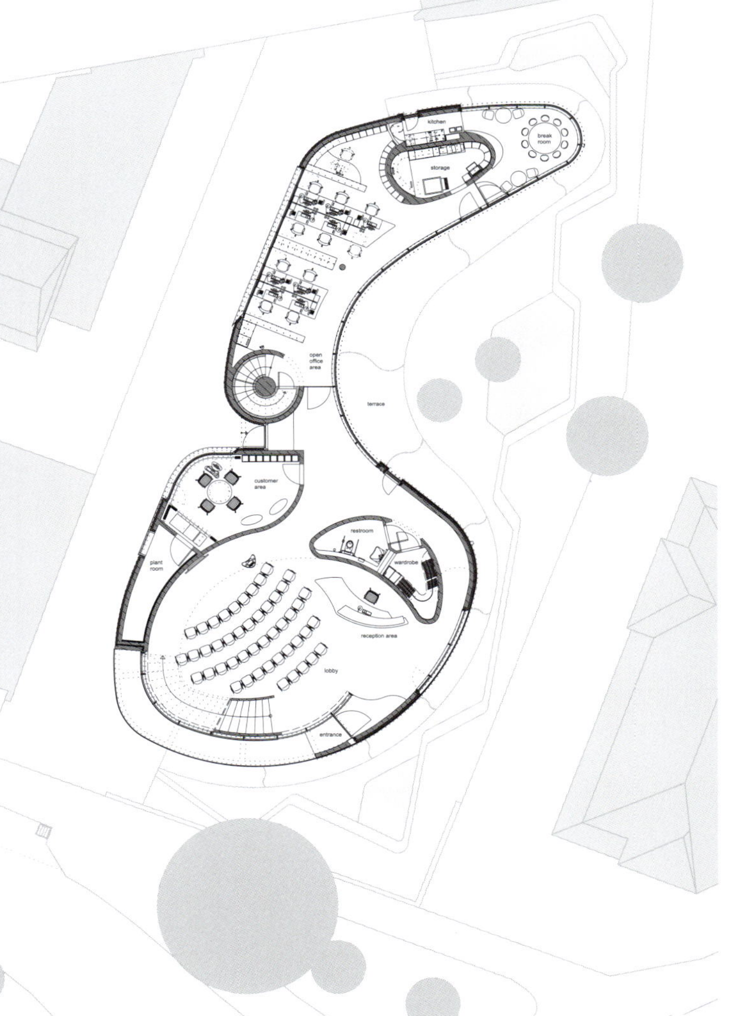

Saegeling Medizintechnik | Heidenau

ground floor

基于周边现有的小规模建筑结构，GERD PRIEBE 建筑设计及咨询事务所设计了一座双层建筑。该建筑曲线优雅，其独具匠心的别致形态和谐地融入了现有的城市规划之中。策划伊始，可持续发展的方方面面就已被纳入考虑之中，并且在设计中发挥了不可或缺的作用。

建筑的整体设计、使用舒适度、视觉舒适度、听觉舒适度以及温度的舒适度都细致地结合在了一起。

新建筑的东面彰显了其宽敞、露天的特点，曲面的玻璃外墙围合出一片僻静的户外区域，看上去像是与室内空间融合在了一起。玻璃外墙纵向延伸了空间的宽敞感，使室内采光充足。建筑分为两个功能区：南面的接待区，或称为“公共区域”，可引领来访者穿过一个双层的休息区，同时该区域也可以作为议事厅使用；休息区的旁边，一层是为客户准备的会议室，二层则是画廊。

玻璃、钢材和混凝土被用作主要建材，目的是保证建筑结构的整体性。该建筑的曲线外形提升了承重墙的坚固性和稳定性，同时还节约了用料，减小了墙的厚度。

外墙上精致的白色灰泥表层与鲜艳夺目、凹凸有致的浅绿色玻璃窗交相辉映。高度可达 5.3 米的弧线形夹层玻璃是在芬兰为该项目量身定做的。这种玻璃首次被用于德国的建筑中，在垂直方向上，该玻璃无需接缝即可密封，横向上则由一个保险条板固定住。许多半径不同的个性的曲面组合使曲面玻璃外壁更加与众不同。外部区域的玻璃窗格至多有两个半径，而内部区域的玻璃窗格最多则是由三个半径组合而成。

为了营造一个高效的办公环境，该设计不再严格区分工作区和活动区，进而创造出了一个流动、高效、实际无支撑的建筑。

Designer Enrico Taranta, Giorgio Radojkovic, Juriaan Calis

RED TOWN OFFICE

Location
Red Town Sculpture Park, Shanghai, China

Area
120m^2

Photographer
Shen Qiang from Shen Photo

Shanghai based practice Taranta Creations designed their own studio located in Shanghai, China.

The design is a reflection of the ongoing creative process within the studio. The intervention seeks to provide an adaptable space that supports a range of informal functions.

The office is situated in a former metal factory. The existing diagonal steel structure was causing difficulties to fulfill the wish of the studio to create two floors. The distance between the structure and ceiling was too less for a traditional office layout of floor and tables. Therefore a floor is created just above the steel structure. Four working stations are placed in the space between the steel profiles. In this way the floor is transformed to one continuous desk, while the four recessed stations provide a more traditional workspace. The large "work floor" invites the designers to use the open space for thinking, sketching, meeting, drafting, modeling, sitting and relaxing. This informal interpretation of office space encourages cross-pollination between different projects and disciplines occurring within the studio.

On the lower floor the individual workstations are placed along the window. A green sculptural table can be used for communal activities. Informal and contoured, the central staircase is reminiscent of a large droplet of water ready to fall from the ceiling. Upon entering the stairway, a highly saturated environment of bright red engulfs and surrounds the individual, starkly marking the transition between the contrasting office areas. The color is repeated on the second floor as an accent to highlight the recessed work areas.

这是 Taranta 创作工作室为自己设计的位于中国上海的办公室。

设计反映了工作室的创意进程。一系列的非正式功能使空间的适应性极强。

工作室所在地的前身是一个五金工厂。由于成对角线状的钢构造的存在，想要建造两个楼层十分困难。对角线状构造和天花板之间的距离很短，不能像传统办公室一样摆放桌椅，于是，设计师就在钢构造之上建了一层楼。在钢构造之间有四个工作台，这样，这一层就变成了一个连续的办公桌，而那四个嵌入式工作台则提供了更规范传统的工作空间。巨大的工作平台供设计师们思考、画图、开会、建造模型和休息放松。对该办公室的非正式诠释促进了不同学科和项目之间的交流。

一层的个人工作台沿窗分布。绿色雕花桌用于公共活动。中央随意的波状楼梯就像一滴快从天花板上坠落的水滴，楼梯通道的内壁是给人以高度饱和感的大红色，围绕着个人工作台，在对比强烈的办公区域形成了很好的过渡。红色在二层的重复使用突出了嵌入式的工作区。

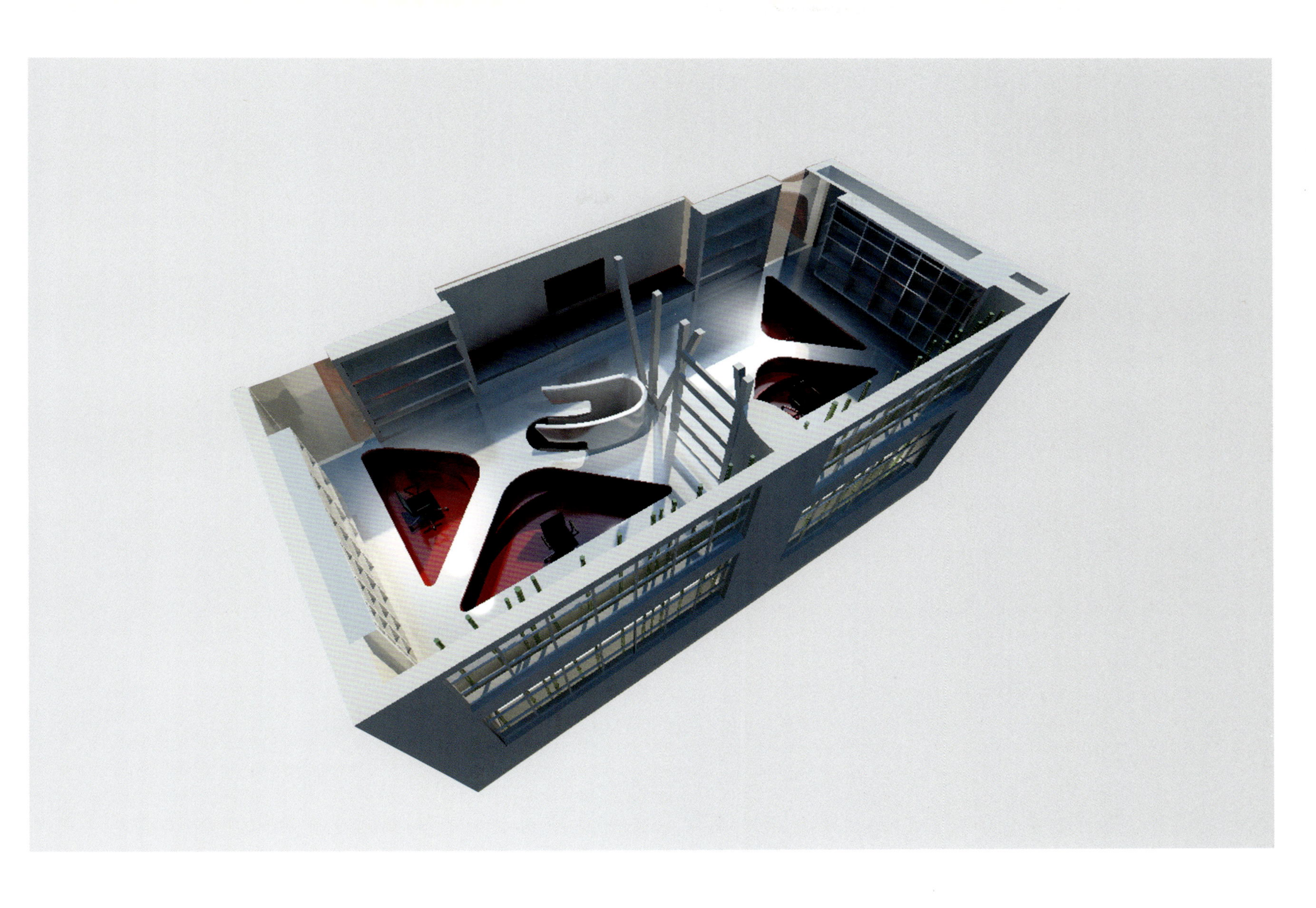

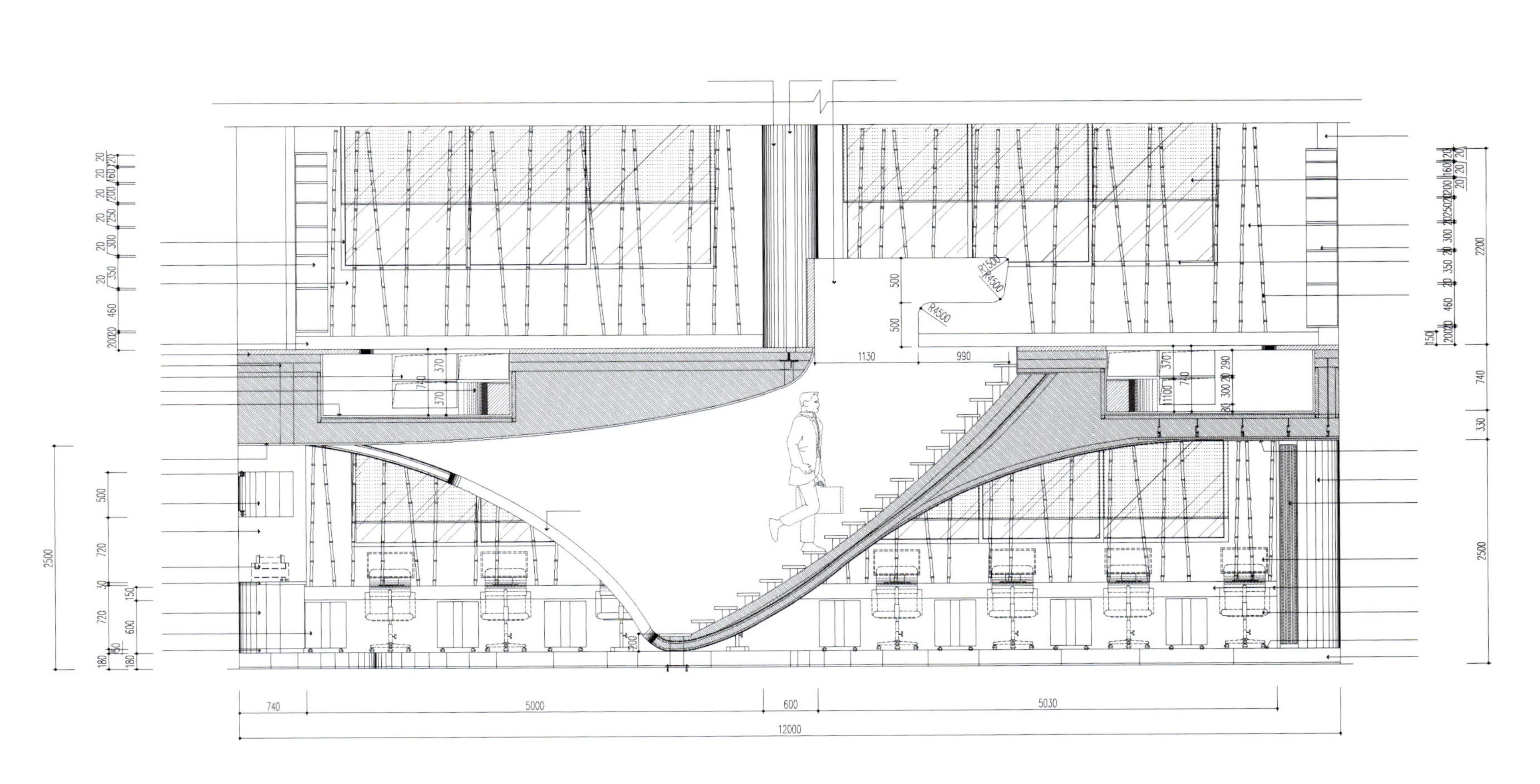

Designer Jose Ramon Madayag Mercado

ASTRA ZENECA THAILAND CORPORATE OFFICE

Design Company
Design Worldwide Partnership

Location
19th Floor Asia Center, Bangkok, Thailand

Area
1,400m^2

FOSTERING HEALTHY WORKSPACES

Having thoroughly researched various new offices of Astra Zeneca worldwide and investigating the brand global corporate identity, a key component to win the contract, dwp clearly understood the foremost corporate principle of this one of the world leaders in pharmaceutical medicines, equipment and treatment: namely, "Health Connects Us All". Astra Zeneca was very keen on healthy surroundings, to foster better communication, connectivity, creativity and a sense of community and pride.

Under these guiding principles, planning was central to achieve the desired goals. Strong branding was essential, using the key Astra colors of purple, yellow and violet, setting against a white organic backdrop. A sense of community was achieved by providing a large breakout space, referred to as "The Business Center" for visitors and staff, right in the heart of the office at the reception hall, with the boardroom placed just adjacent. The boardroom was provisioned to opening up onto the breakout space with large, flexible swivel doors.

Another playful key feature of the client space were the ceilings adorned with organic shapes of drop down cove lighting, mimicking the shape of the corporate logo, as well as made-to-order acoustic gypsum boards perforated with holes of different sizes, uniquely creating a beautiful effect, resembling a bowl of medicine tablets.

AstraZeneca

Astra Layout

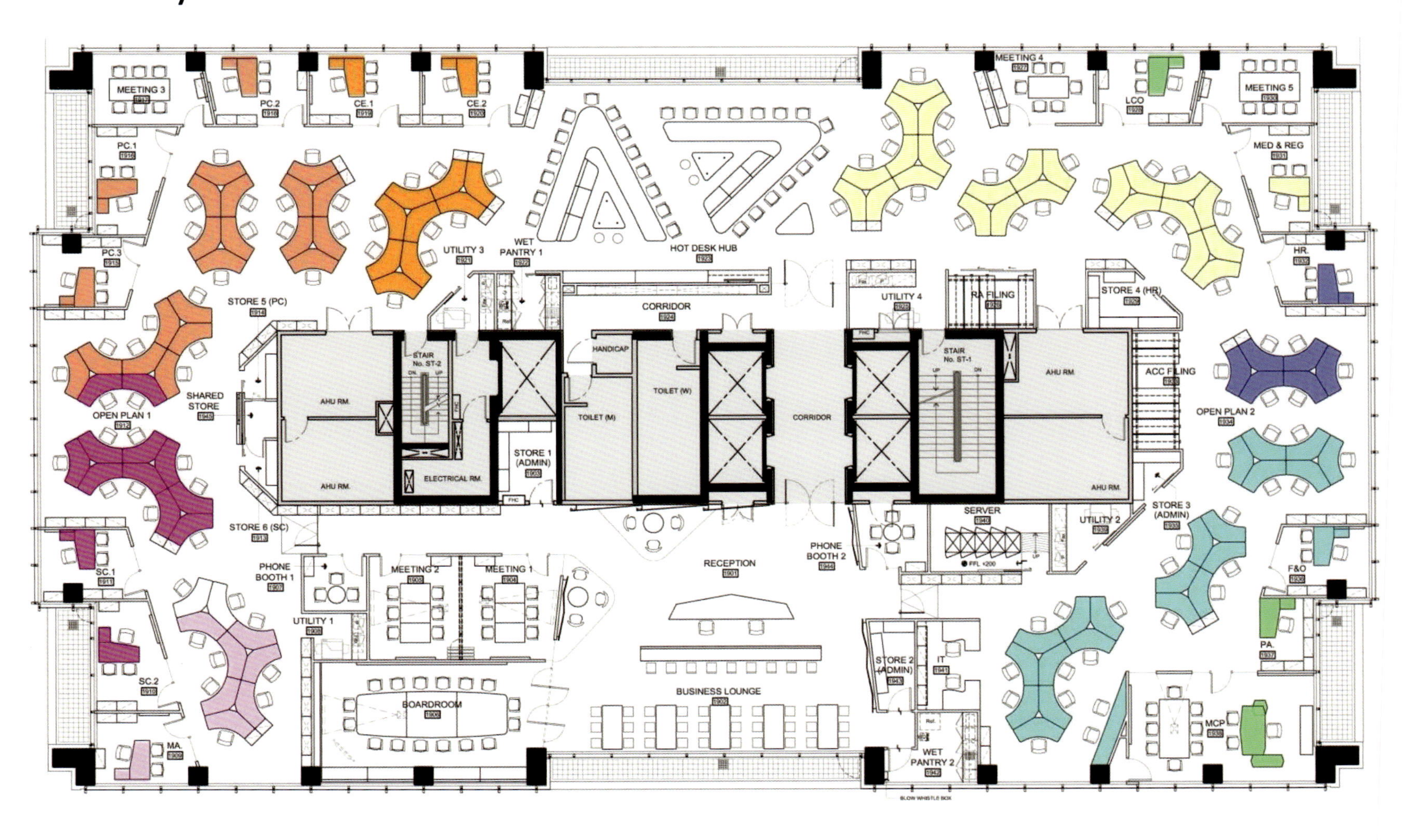

创造健康的工作环境

dwp 拿下这个合同的一个关键因素在于他们深入研究了阿斯利康遍布全球的各个新办公室以及这个全球公司的品牌特征。这家全球医药、医疗设备及医护领域的领军公司的企业原则是：“健康把我们大家联系在一起”。阿斯利康曾经一度热衷于卫生环境建设，以创造更好的沟通性、连通性、群体意识和自豪感。

根据这些指导性准则，本案计划的核心是达到期望目标。阿斯利康的品牌优势至关重要。设计师使用了其主打色紫色、黄色和紫罗兰色，背景色则选择了自然的白色。在接待大厅的正中央为来访者与工作人员设置了一个“商务中心”——一个宽敞的休闲区，以此来创造一种社区感。会议室正对着休闲区敞开，中间是一扇很大的灵活旋转门。

客户区里另一个有趣的特征是天花板，上面装饰有模仿公司商标形状的有机图案，像落入小山坳中的光线，再结合定做的吸声石膏板上凿出的大小不同的孔，创造出别具一格的完美效果，仿若一碗药片散落在天花板上。

for patients and
stakeholders
great medicines .... because
Health-One of the world's biggest agendas.Bigger than any one thing that we might be working on. The thread that binds us all together in an increasingly collaborative effort to create better healthcare solutions.
Connects-Linking people all around the world who have different roles to play in healthcare as patients, payers, scientists and business interacting and working together.
Us-You and me, our colleagues, our scientific and business partners, customers and other stakeholders. Everybody we work and interact with and whose lives we touch with our science, products and services.
All of us and the communities that we serve internationally in more than 100 established and emerging all working together to achieve common goal of improved
Health
Connects
Us All

FOCUS | 2011
Leadership
Work-life
Integrity

FOCUS | 2011
Leadership
Engagement
Work-life
Visit the intranet for more
Integrity
Have your say and help shape the future of AstraZeneca...
Survey open
5 – 23 September

Have your say and
help shape the future
of AstraZeneca...
Survey open
5 – 23 September

Designer Shafie Latiff, Ben Hwang, Shark Rased, James Loh Shixiong

DDB

Design Company
BBFL

Project Management Team
Cushman Wakefield

Builder Main Contractor
Crown Construction

M&E Main Contractor
Jag Engineering

Location
Pico Center, Singapore

Area
3,716m^2

Photographer
Living Pod

The design process began with observing the designers at DDB work as a diverse group which involves a high level of discussion, debate and intervention sessions amongst each team with a common goal of achieving a creative solution for their clients. Group discussion activity was crucial for the team where workspace becomes a social network area rather than a confined individual space.

The design process has allowed the designers to create a bridge between two disciplines which played a huge role in the concept development of the entire DDB environment creating a conspicuous identity.

The workspace concept that revolves around DDB's office wasn't just about its corporate colors or even the branding of its company but creates a collective space that leans towards the culture of the people in DDB itself. This lead the designers to create spaces in which improves and promotes inter-connectivity between creative individuals. To enhance this notion, spaces within the work environment weren't crafted out with walls or any form of concealed demarcation, compositing to an open office setting.

Territories are divided and screened by discussion and collaborative areas within the office. The aim was to enhance the current culture and to encourage a more lively work space, which will ultimately improve a better work flow. The workspaces are designed as long communal tables supporting an open office environment. The designers design spaces that allow these communal areas to be part of the workspace. Discussion bars are dispersed within team work areas. Even libraries and the staff pantry spanning across both floors were used as collaborative spaces for discussions.

Conference rooms were meant to be flexible and open where staff would frequently occupy for internal presentations and brain-storming sessions. Gym-like staggered benches were introduced for audiences to participate in these sessions. Arrival lobbies on the other floors were treated with interactive projection of the collection of works that inspired DDB and a source of expression for the staff.

Recalls Shafie Latiff during the development stage of DDB's office design, "To sculpt an environment for designers, it is essential that the spaces are open, flexible and inspiring where staff could express themselves freely just like a white open canvas."

DDB
EXIT

EXIT

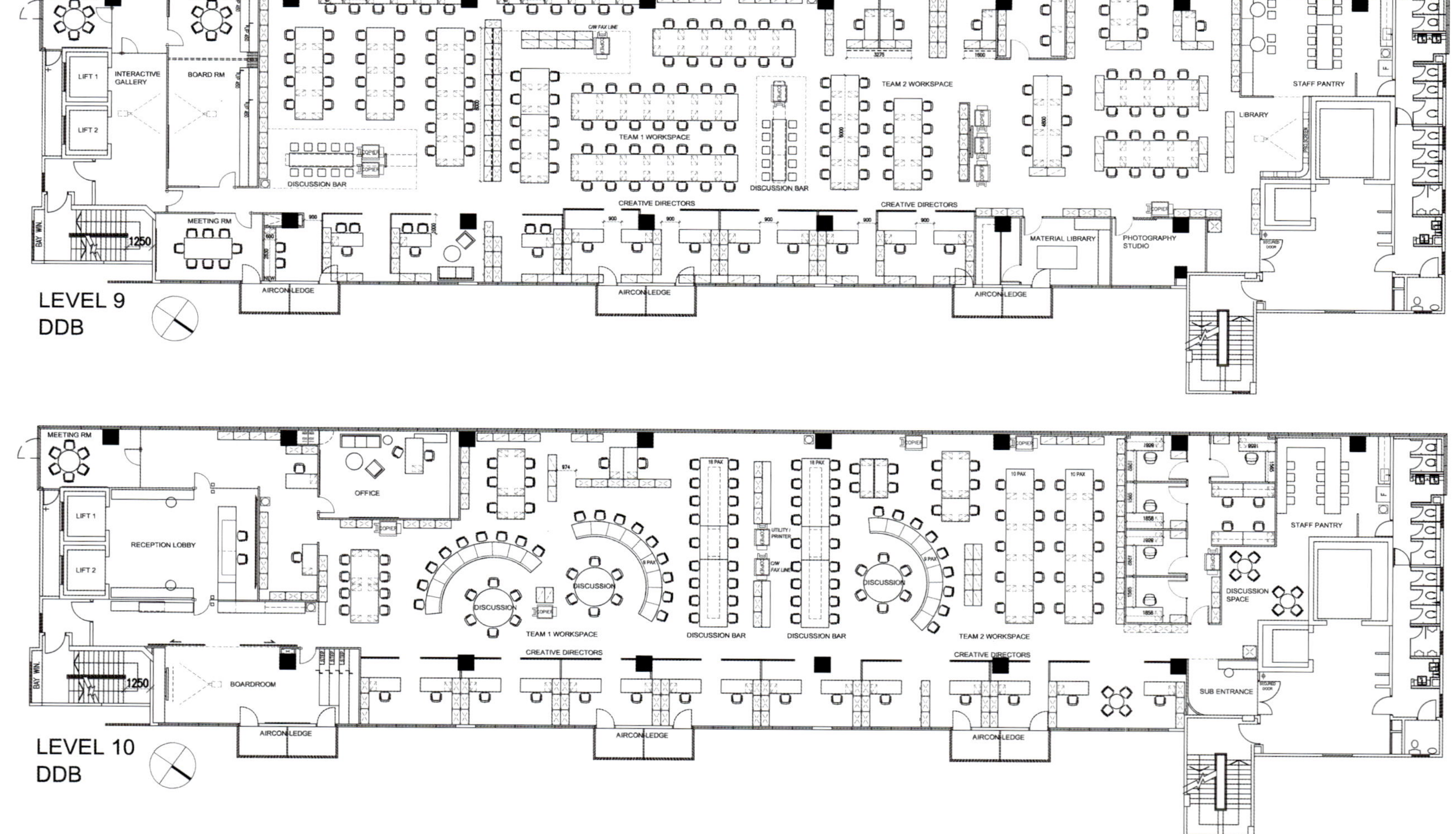
MEETING RM
MEETING RM
LIFT 1
LIFT 2
INTERACTIVE GALLERY
BOARD RM
DISCUSSION BAR
TEAM 1 WORKSPACE
DISCUSSION BAR
TEAM 2 WORKSPACE
STAFF PANTRY
LIBRARY
CREATIVE DIRECTORS
CREATIVE DIRECTORS
MEETING RM
MATERIAL LIBRARY
PHOTOGRAPHY STUDIO
AIRCON LEDGE
AIRCON LEDGE
AIRCON LEDGE
LEVEL 9
DDB
MEETING RM
LIFT 1
LIFT 2
OFFICE
RECEPTION LOBBY
DISCUSSION
DISCUSSION
DISCUSSION
TEAM 1 WORKSPACE
DISCUSSION BAR
DISCUSSION BAR
TEAM 2 WORKSPACE
STAFF PANTRY
DISCUSSION SPACE
CREATIVE DIRECTORS
CREATIVE DIRECTORS
BOARDROOM
SUB ENTRANCE
AIRCON LEDGE
AIRCON LEDGE
AIRCON LEDGE
LEVEL 10
DDB

DDB

RAPP

GOLD
SILVER
Cannes
Effies
Spikes
GOLD SILVER BRONZE
CANNES
SOAA
H.O.F
SPIKES
INK!
EFFIES
ONE SHOW
D&AD
ADFEST
CCA

IMAGINATION WILL TAKE YOU EVERYWHERE
B. IMAGINATION WILL TAKE YOU EVERYWHERE.
FROM A TO BEYOND
In a business that's all about metrics and data, it's easy to forget we're moved by emotion. 'From A to Beyond' is a daily reminder to take that leap of imagination. We want to step in to work every morning and be inspired by something new, different, exciting. How else can we move others?

03-28
FLORENCE RODRIGUEZ, 27
MAID
03-26
SENG WEK THYE, 67
YONG TAU FOO SELLER
03-24
CHONG LEE ANG, 35
03-22
02-28
GABRIEL LEE, 21
UNIVERSITY STUDENT
02-26
LIM SEOW TECK, 68
RETIREE
02-24
EKISHA GANESH, 48
02-22
DIANA SEOW, 15
00-00
JUNE YAP, 23
01-24
01-26
MANDY WONG, 17
STUDENT + PART-TIME WORKER
01-28
TAKE ONE

TYPOGRAPHY 10
GRAPHIS DESIGN
CONCISE WORLD ATLAS

该设计过程由观察开始：恒美广告的设计师们能够组成多样化的团队，在组内进行讨论、争辩，互相协调，目的是为客户提供一个独具创意的解决方案。对于团队而言，小组讨论十分必要，因此，他们的工作场所更像是个社交空间，而不是有限的私人空间。

在设计的过程中，设计师在两种不同的准则之间搭建起一座桥梁，这为恒美广告公司营造个性鲜明的整体环境起到了至关重要的作用。

恒美广告公司办公室的设计理念不仅与公司的颜色或商标紧密相关，更要创建一个贴近恒美人文化的公共场所。因此，设计理念引导设计师们创造出一个有利于内部交流的空间，能够加强有创造力的员工之间的交流。为了强化这一理念，工作区域内没有设立墙面或其他任何隐形界限，而是设计成一个开放式的办公环境。

办公室内的区域被分隔成讨论区和协作区，目的是强化现有的公司文化，建立一个更活跃的工作场所，进而从根本上改观现有的工作流程。工作区设计了许多长桌来打造开放式的办公环境。设计师的设计使这些公共区域都可以成为工作场所的一部分。讨论吧分散在团队合作区内，甚至连横跨两层的图书室和员工食品储藏室都可以用作共同讨论区。

会议室设计得灵活而开放，员工可以经常来这里做内部演示或进行头脑风暴（创造力的集体训练）。像体育馆式错列排放的长椅使观众可以参与到讨论中去。位于其他楼层的入口大厅设有交互式投影仪，播放着恒美已完成的作品，这不仅能够鼓舞恒美的员工，同时也是员工自我展示的一种方式。

回忆恒美广告公司设计的阶段，莎菲宜·拉提夫说："为设计师设计工作环境时，空间的开放性、灵活性和启发性是很重要的。就如一块展开的白色帆布，能够激发员工自由地表达自己的想法。"

Designer Dean Lah, Milan Tomac, Anže Zalaznik, Jure Kocuvan

NKBM

Design Company
Enota

Structural Engineering
G-biro Gecev

Mechanical Services
Biro ES

Electrical Planning
Biro ES

Location
Ljubljana, Slovenia

Area
1,150m^2

Photographer
Blaž Budja

The interior design project for the NKBM Bank branch utilises a territorializing system that was developed as a set of basic functional elements. They regulate the spatial ratio between customer and employee areas and provide private spaces for banking discussion.
The conventional vertical screen mutated into a deformed glass shell of a complex geometry in order to provide optimal spatial ergonomics in a limited amount of space. Using a computer modelling algorithm, the geometry was triangulated in order to simplify production. Glass triangles are locally assembled with steel clips, which enable angle adjustments for individual fixation. Glass shells are a self-supporting system that doesn't require additional structure.
Variable opacity is provided with a geometrical print of three different densities on the inside of the shell; organic patterns on the exterior visually unify the screen set.

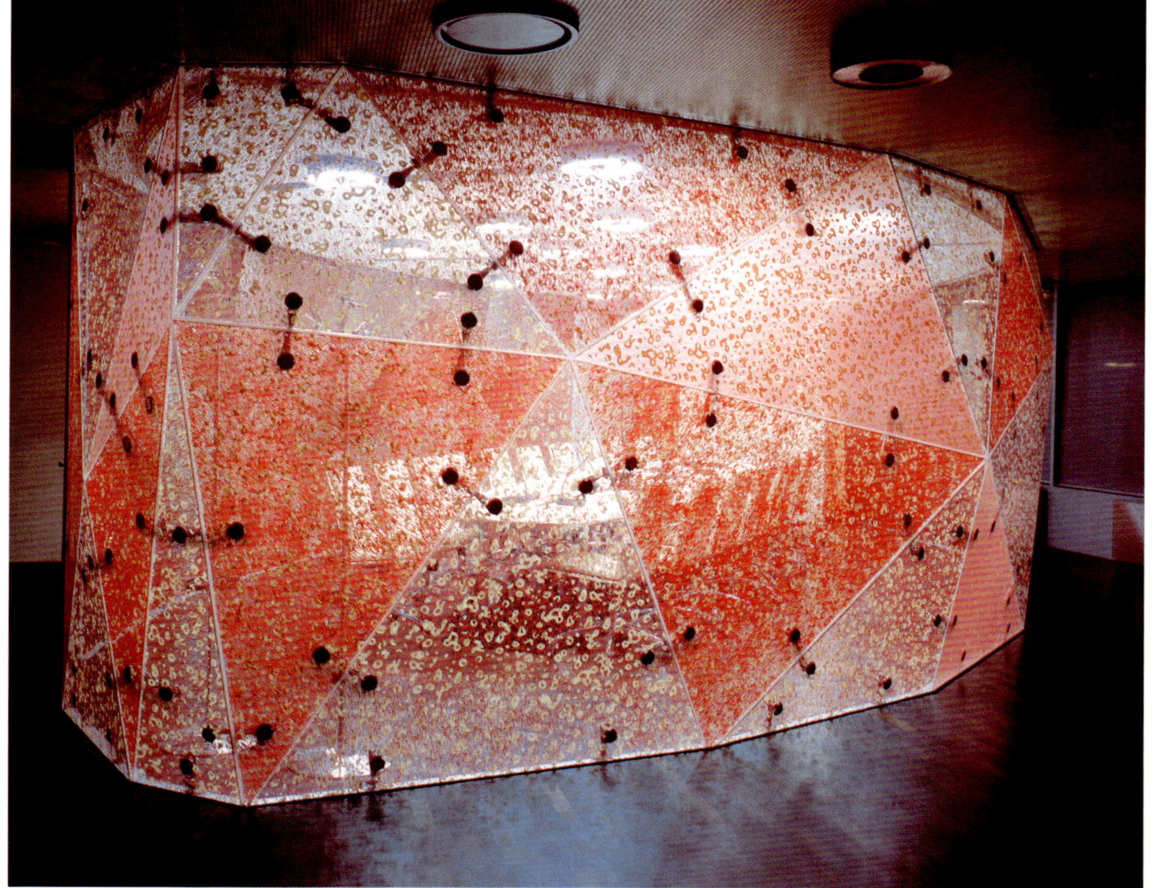

NKBM 银行分行的室内设计项目运用了一个由一系列基本功能元素发展而成的区域化体系，规划了顾客和员工区域的空间比例，并提供了讨论银行业务的私密空间。

为了在有限的空间内体现出对人类工程学的最佳运用，本案将常规的垂直屏幕转变为呈复杂几何形状的变形玻璃外壳。为简化施工，设计师运用电脑模拟运算，将几何图形分解成三角形，并在现场用铁夹安装玻璃三角，这样可以根据个人喜好来调整角度。玻璃外壳是一个无需附加结构而能够自我支撑的系统。

外壳内部有三种不同浓度的几何印迹，达到了可变不透明的效果，其外部的组合图案使外屏达到了视觉上的统一。

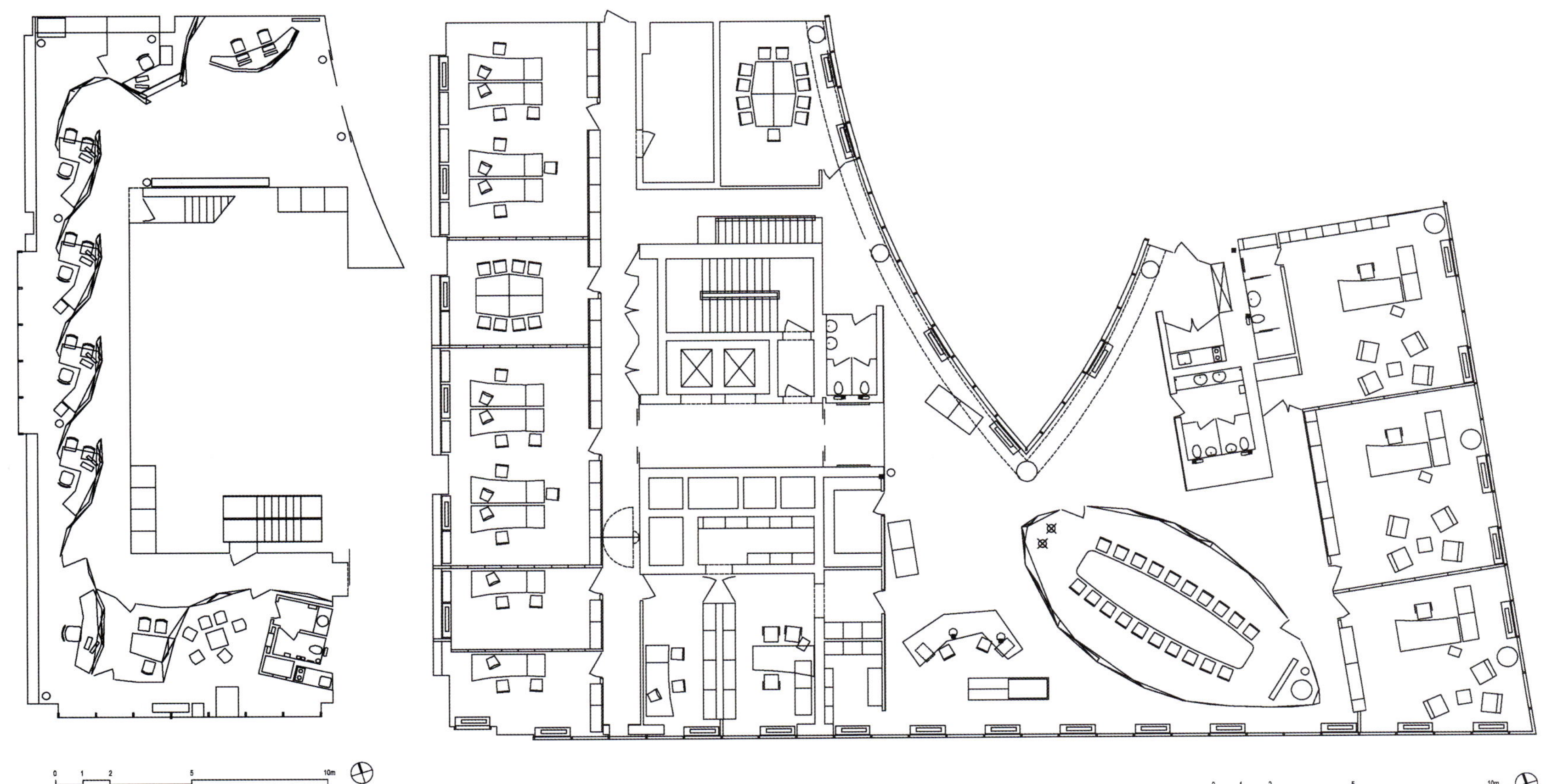
0
1
2
5
10m
01 nkbm ground floor plan (1:200)
0
1
2
5
10m
02 nkbm first floor plan (1:200)

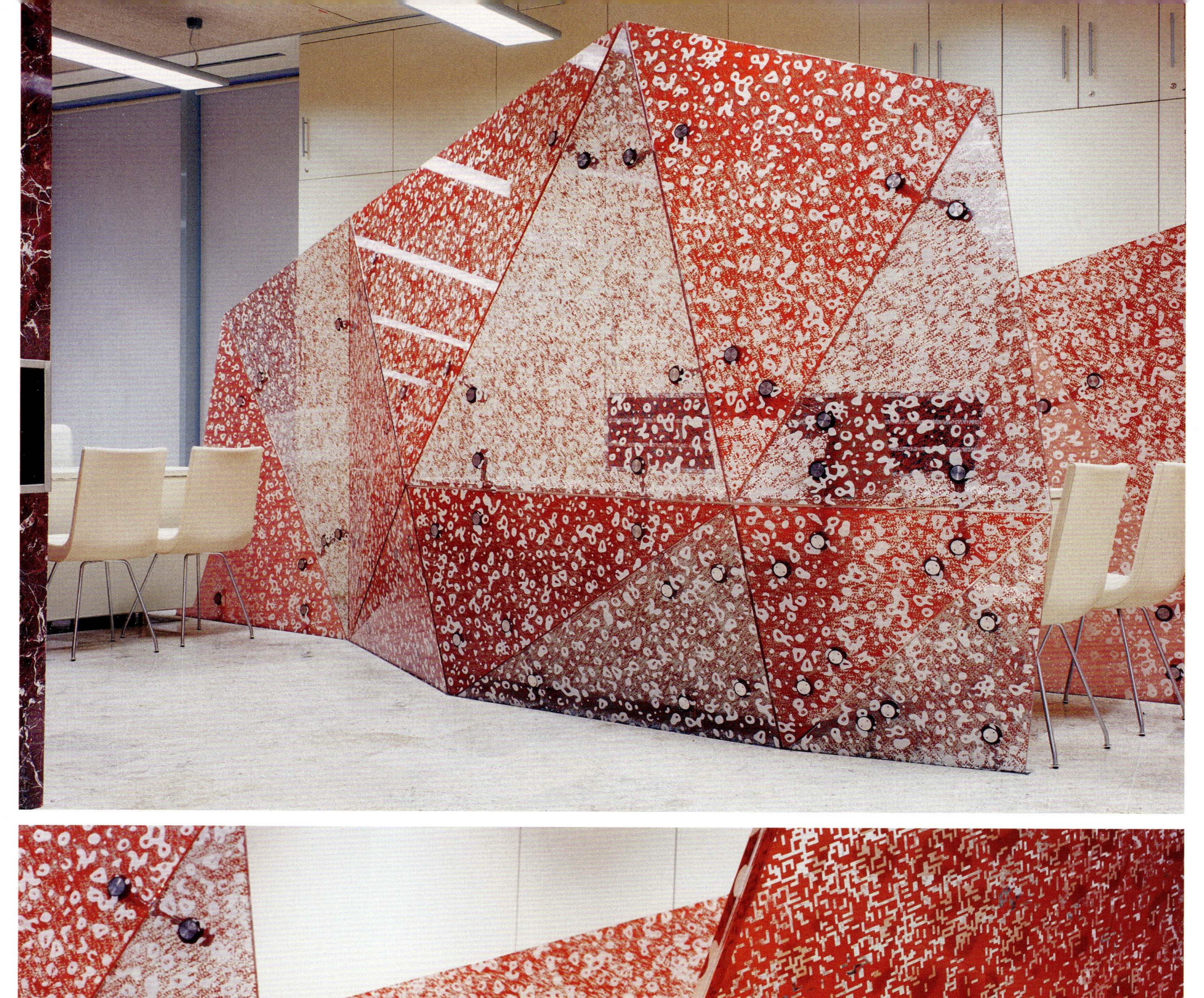

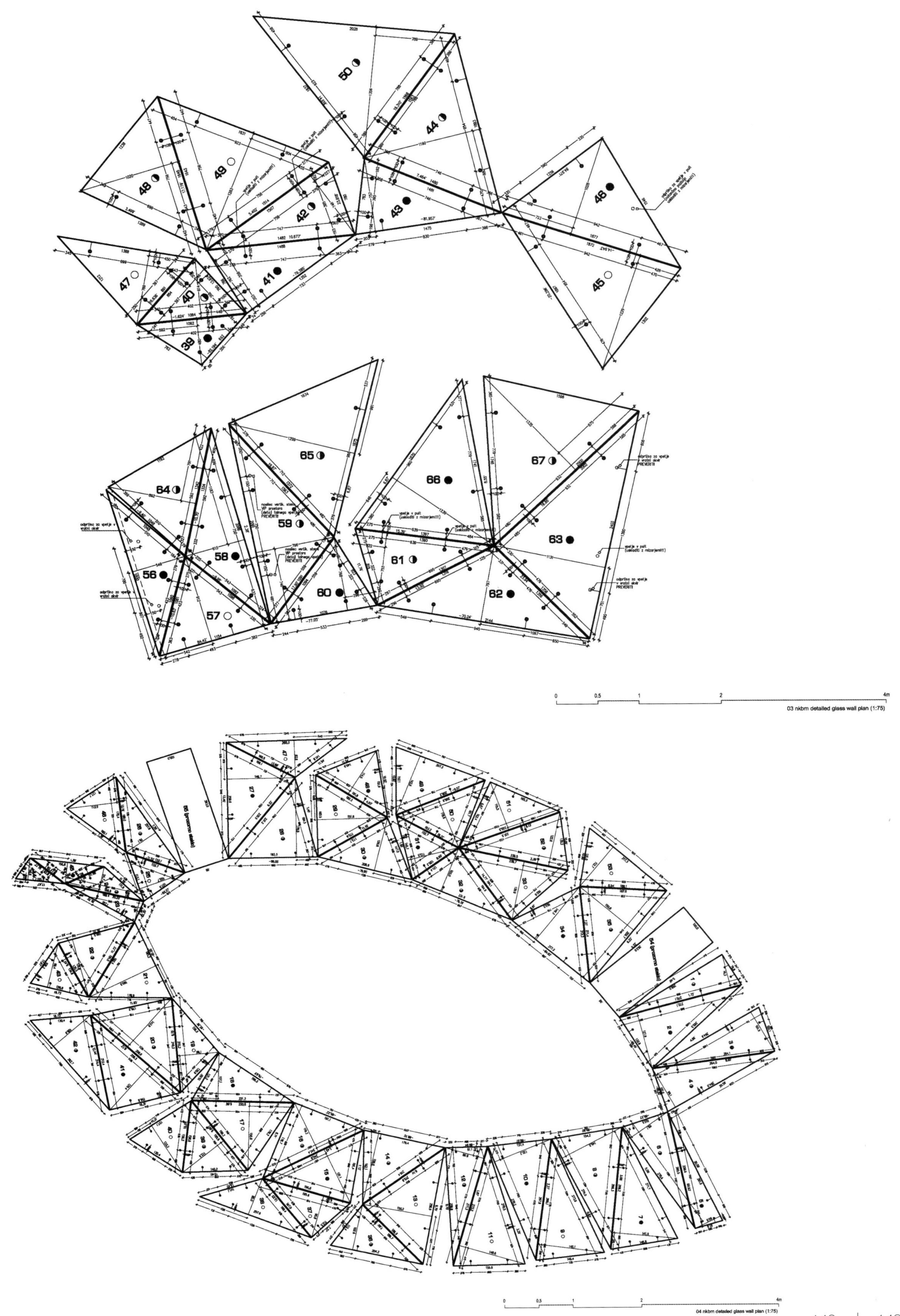

03 nkbm detailed glass wall plan (1:75)

04 nkbm detailed glass wall plan (1:75)

Designer Mike Atkin, Mark Panckhurst, Robert Weller, Steven Choi, Alex Jiaravanont

PHOENIX TV

Design Company
Head Architecture

Location
Hong Kong, China

Area
9,335m^2

Positioned in the South East corner of the Tai Po Industrial Estate, the building faces directly West. The building comprises of two wings with a central circulation core housing reception and entrance. The studio placed in the larger southern section takes up the ground floor and new mezzanine of the available 3 storeys.

PSTV needed to move out of their shared Headquarters (with Star TV) in Hung Hom, requiring a studio within Hong Kong that could compete with large international news organizations in Asia, Europe and the US.

The studio includes a main News Area comprising four news desks, video/audio mixing areas, presenter suites, graphic design office and a data center. The above and journalists were to be visible on camera. An additional six independent studios ranging from 180 m^2 to 480 m^2 were also constructed. A VIP walkway links all the major areas at Mezzanine level. This allows investors and dignitaries to walk through the working studio.

In order to accommodate the studio, one internal structural column was removed and four quadrants of floor slab and beam were removed to create an atrium. A mezzanine floor was inserted between Ground and First Floor, and much of this new space faces on camera areas. The VIP walkway was incorporated on this level, running from the gallery around the studio to the editing corridor.

Journalists and associated staff are accommodated throughout the entire floor area seated in a number of sunken "pods" or in rows of fixed seating to provide a dramatic backdrop for the news readers. Lighting around the studio is created by color controlled LED allowing for an infinite variety of color changes.

大楼坐落在香港大埔工业园区的东南角，面朝西。大楼包含两个侧翼，中央流通核心区包含一个接待处和入口。演播室在空间大一些的南面，占用了三层楼中的一层和新夹层。

凤凰卫视要从与星空卫视在红磡共享的总部中搬出来，需要在香港拥有一个可以和亚洲、欧洲或美国的其他国际新闻机构相媲美的演播室。

演播室包括由四个新闻台组成的主新闻区、视音频混合区、平面设计办公室和数据中心，通过摄像机人们可以看到以上区域和新闻记者。除此之外，还有 6 个 180 ~ 480 平方米不等的独立演播室。VIP 通道通向新夹层的所有区域，这样可以方便投资者和重要人物在工作室内自由穿行。

为了可以容纳演播室，设计师拆除了一个内部结构圆柱和 4 扇楼板，横梁也被改造成门廊。在一层和二层之间还插入了一个夹层，夹层的大部分都对着摄像区域。VIP 通道也在这层，环绕着演播室从边座延伸至走廊。

新闻记者和相关工作人员分布在楼层各处凹陷的“豆荚”中或者一排排的固定座位里，以便为新闻播报员提供一个完美的背景。环绕着演播室的可控 LED 灯可以满足多样的色彩变化。

1.4105/ 1.4106
0.6439/ 0.6444
國際專列

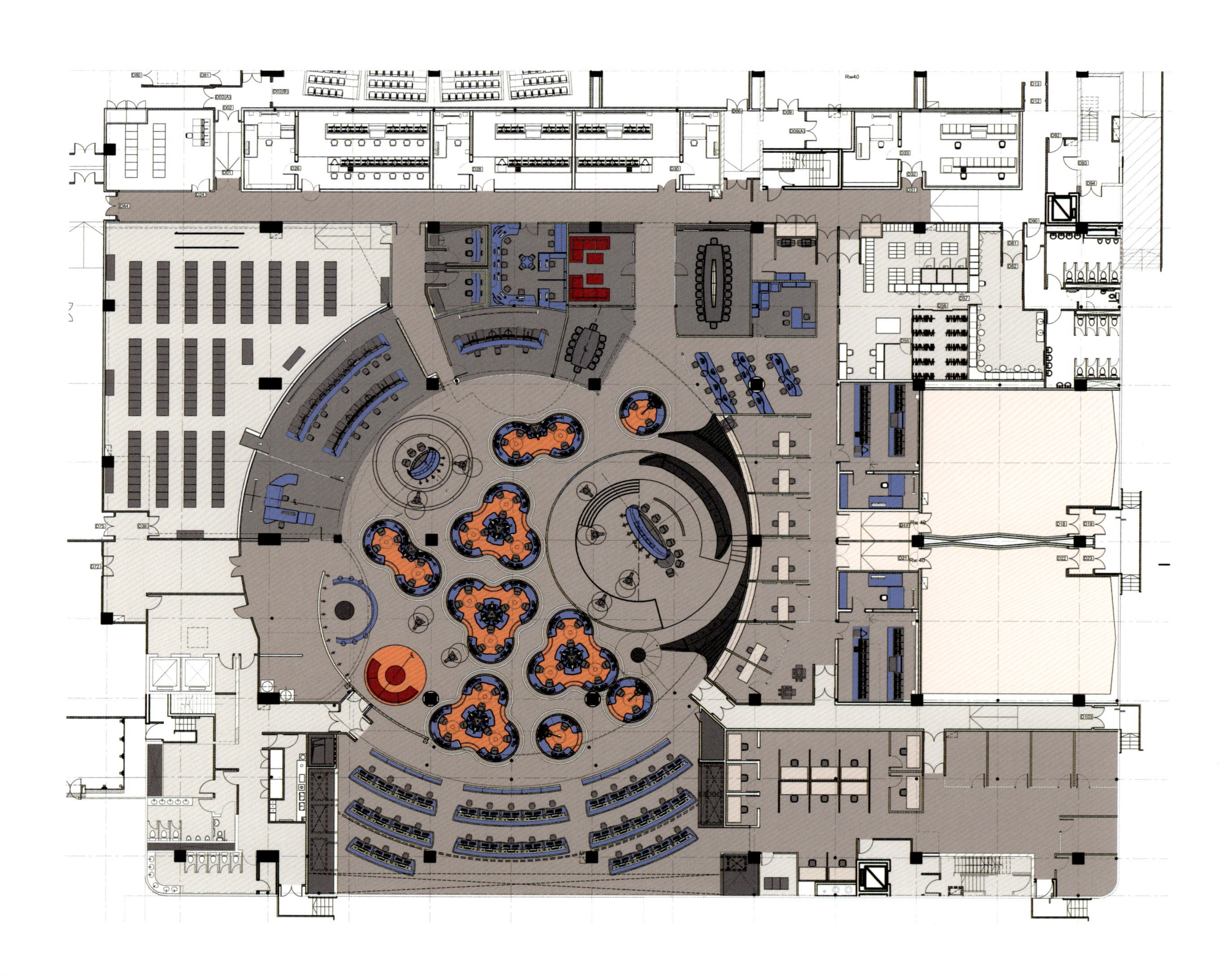

Beijing 1
Beijing 2
Tai Pei 1
Tai Pei 2
Shanghai
18:49:36

CAM 1
CAM 2
CAM 3
CAM 4
EMG OUT
19:01:08
PIPE 1
PIPE 2
REM 2
OM4
REM 1
OM3
CAM 9
CAM 10
DY B+
DY A+
REM 6
REM 5
OM2
OM1
DY B
REM 4
REM 3
DY A
REM 10
REM 9
REM 8
REM 7
AUTOCUE
S_RD OP
DY D
PGM
PVW
SI CAM1
SI CAM2
DY F
Panasonic
DIGITAL D-12

Designer Ippolito Fleitz Group

AGENCY BRUCE B./EMMY B.

Location
West Stuttgart, Germany

Area
500m^2

Photographer
Zooey Braun

Bruce B. /Emmy B. is a prestigious design agency that specializes in communication design and events. The aim in designing a suitable interior for this agency was to faithfully translate what epitomizes their work into architecture. Objects and design elements initially appear to be in diametric opposition, yet strike a harmonious balance throughout. They are emblematic for the wish of Bruce B./Emmy B. to respond to their clients with an open, strategic mindset and an excellent command of antithetical thinking.

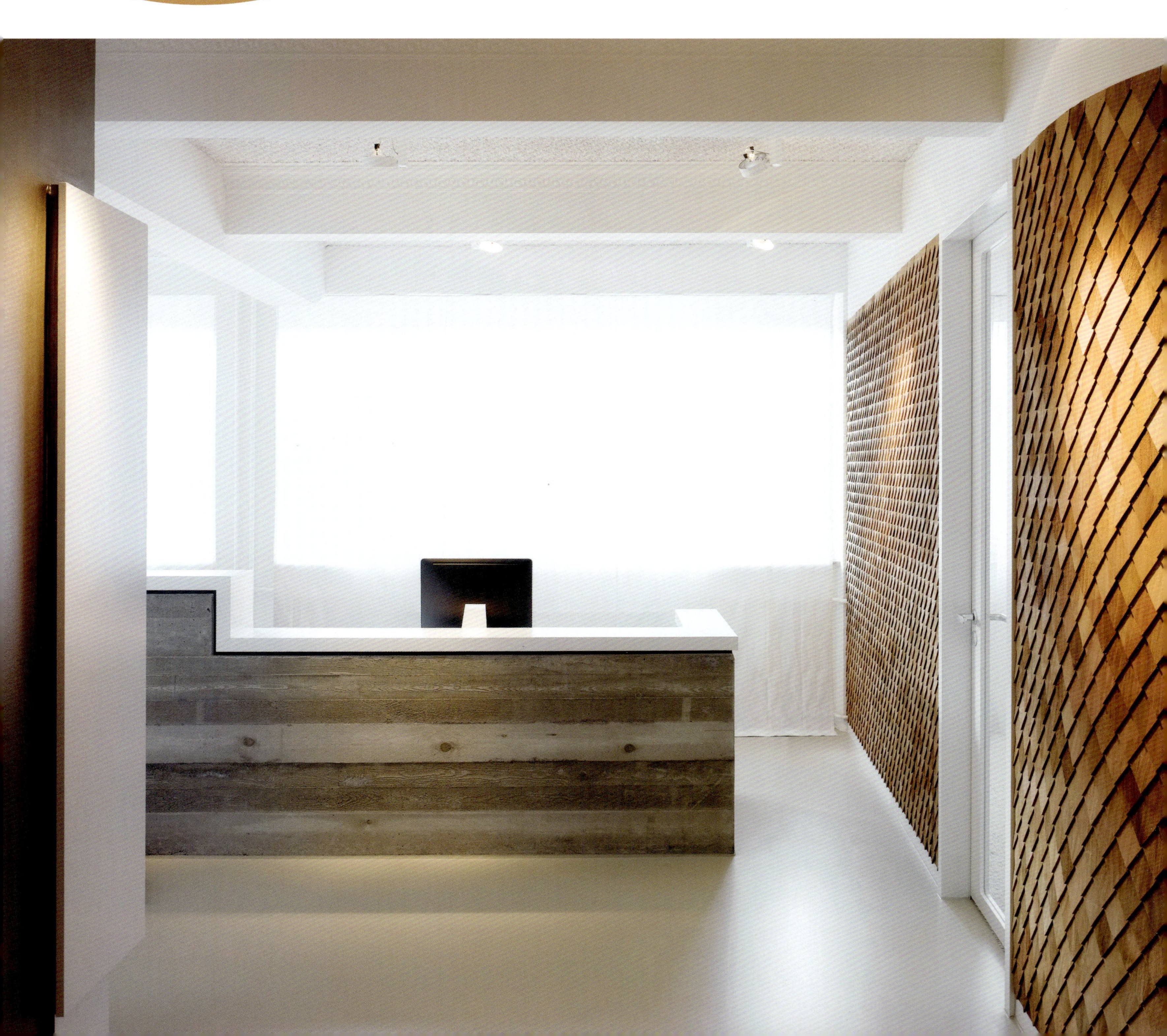

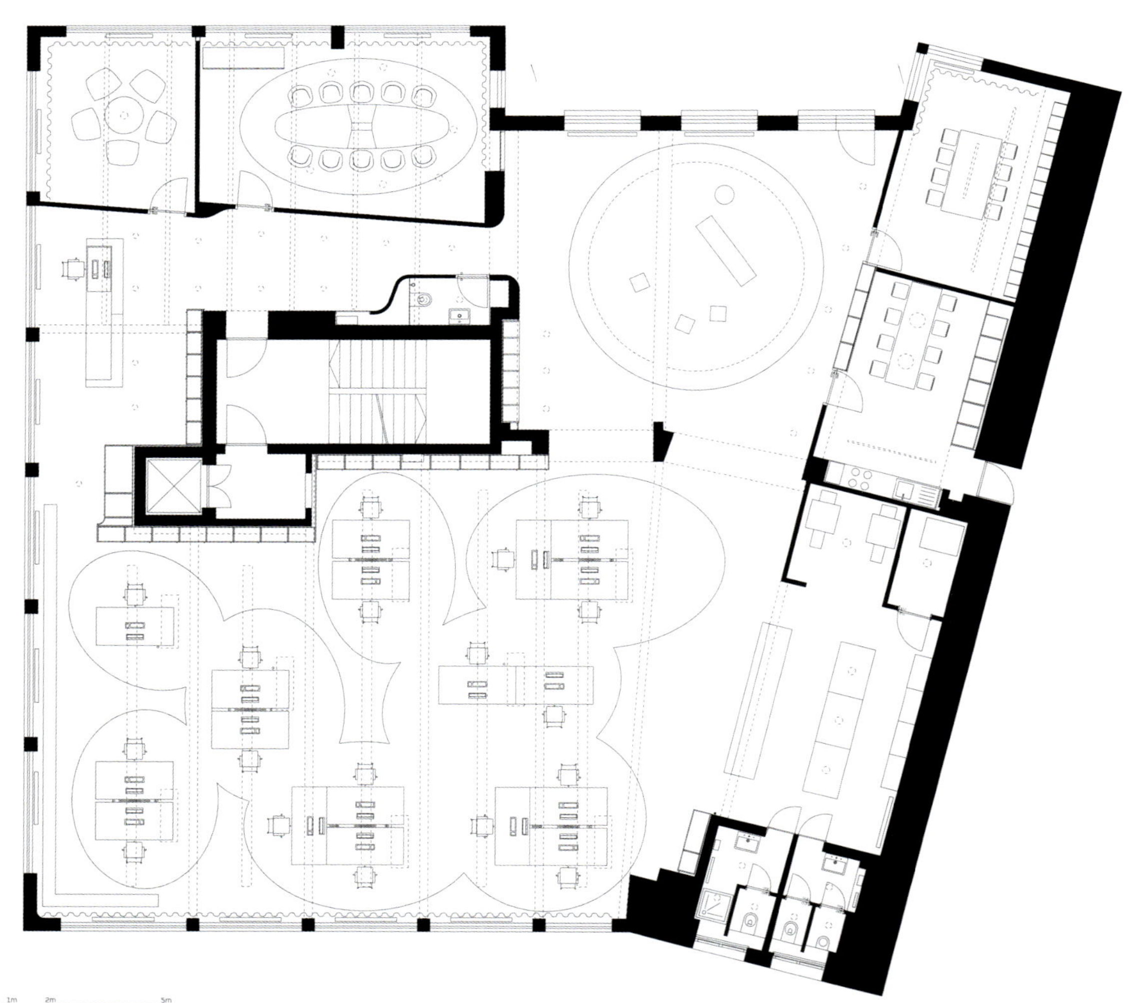
0
1m
2m
5m

Bruce B./Emmy B.是一家久负盛名的设计机构，主营通讯设计和事务。此次室内设计的目的是将这家机构的工作性质通过建筑形式真实地表现出来。起初，设计师觉得构思出的设计元素表现出来的效果与设计的目的背道而驰，但后来却发现，其实二者在各个方面都相辅相成，达到了完美的平衡。这样的设计代表了Bruce B./Emmy B.的初衷——向客户展现他们开放、富有谋略的工作精神和优秀的反向思维能力。

Designer Anton Mossine, Olesia Sokolova, Stas Kirichenko

WHITE SQUARE OFFICE

Design Company
Mossine & Partners

Location
5A "White Square", Lesnaya Street, Moscow, Russia

Area
430m^2

Photographer
Alexei Naroditsky

The office space for the financial company "Xenon" is designed by Anton Mosin's architectural studio.

The office was considered as a cell – a perfectly functional unit. Its structure was realized in the working scheme: the architects allocated the various zones which are responsible for different functions, and correlated them with those of a cell. An external membrane of the office space is a wall with big windows, which transmits a lot of light. The most noisy and active elements (trader's office, kitchen and drivers' room) are concluded in an impenetrable capsule of a nucleus and isolated from other space with soundproofing ceiling. The meeting rooms are like mitochondria – the power centers of a cell. Their membrane is similar to an external one – almost pellucid. But, if necessary, glass capsules can lose its transparency – "rain" curtains which slide out along the perimeter of glass partitions on special curtain rails are provided for this purpose. In the free space of the office analytics perform their part of the work. Like the ribosomes which synthesize proteins, they are responsible for the viability of the whole organism.

All these functional elements and zones represent the centers of different processes, which are providing a metabolism and energy exchange in this cell-office.

conception

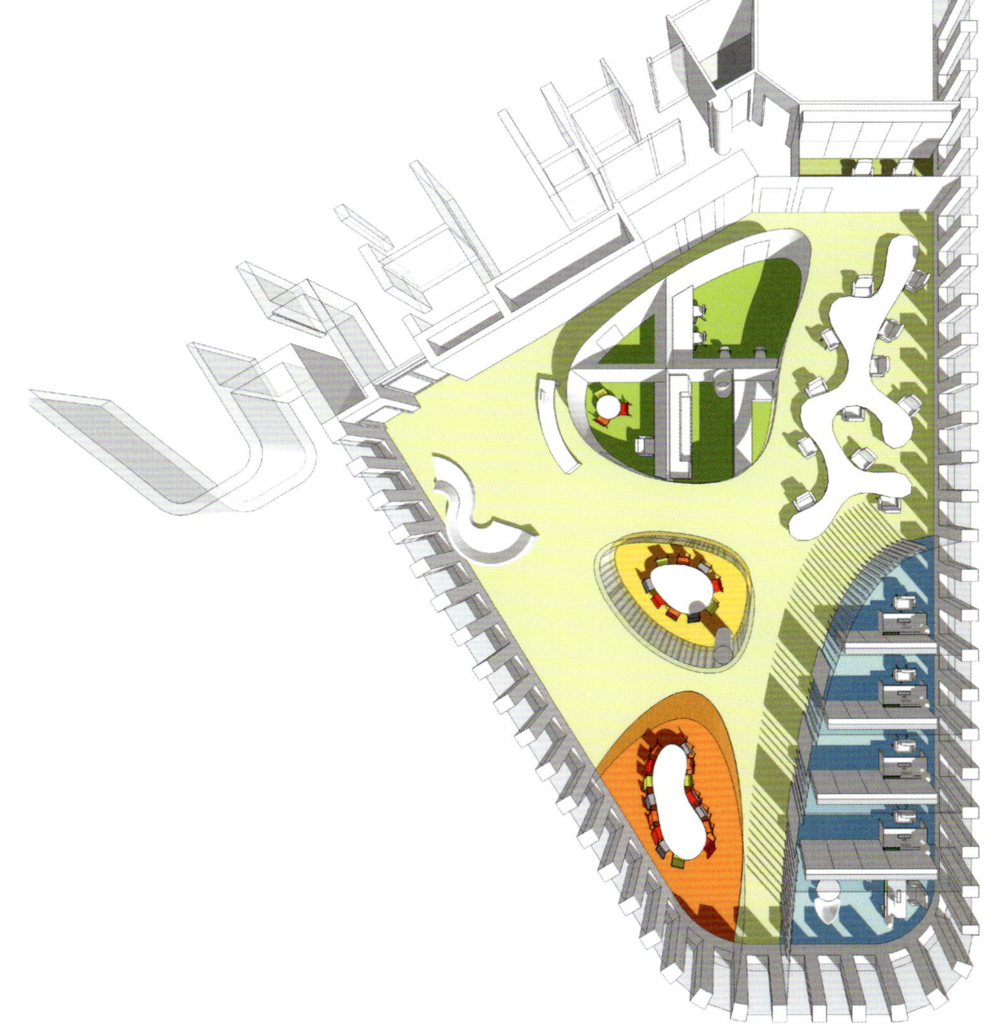

金融公司“Xenon”的办公场所由Anton Mosin的建筑事务所设计完成。

该办公场所被比作一个细胞，即一个功能强大的个体。其结构在实施方案中得到了体现：建筑师将多种功能不同的区域进行分配，再将它们彼此关联起来。一面由巨大玻璃铺成的墙体充当了建筑外部细胞膜的角色，这些玻璃墙可透入大量光线。最热闹活跃的元素（业务员办公室、厨房和司机办公室）被置于细胞核不透光的胶囊里，彼此之间由隔音板隔开。会议室则像是线粒体——细胞的能量中心。会议室的细胞膜与外部的类似，几乎是透明的。但必要时，玻璃胶囊也可以失去透明性——特制窗帘围栏上可以沿玻璃隔墙周边滑动的“雨”帘就是为此而设。对于办公室内的自由空间，解析学发挥了作用。就像合成蛋白质的核糖体一样，它们负责整个有机体的生存能力。

所有这些功能性元素和区域都代表不同活动的核心，而正是这些活动进程保证这座细胞办公室持续进行新陈代谢和能量转换。

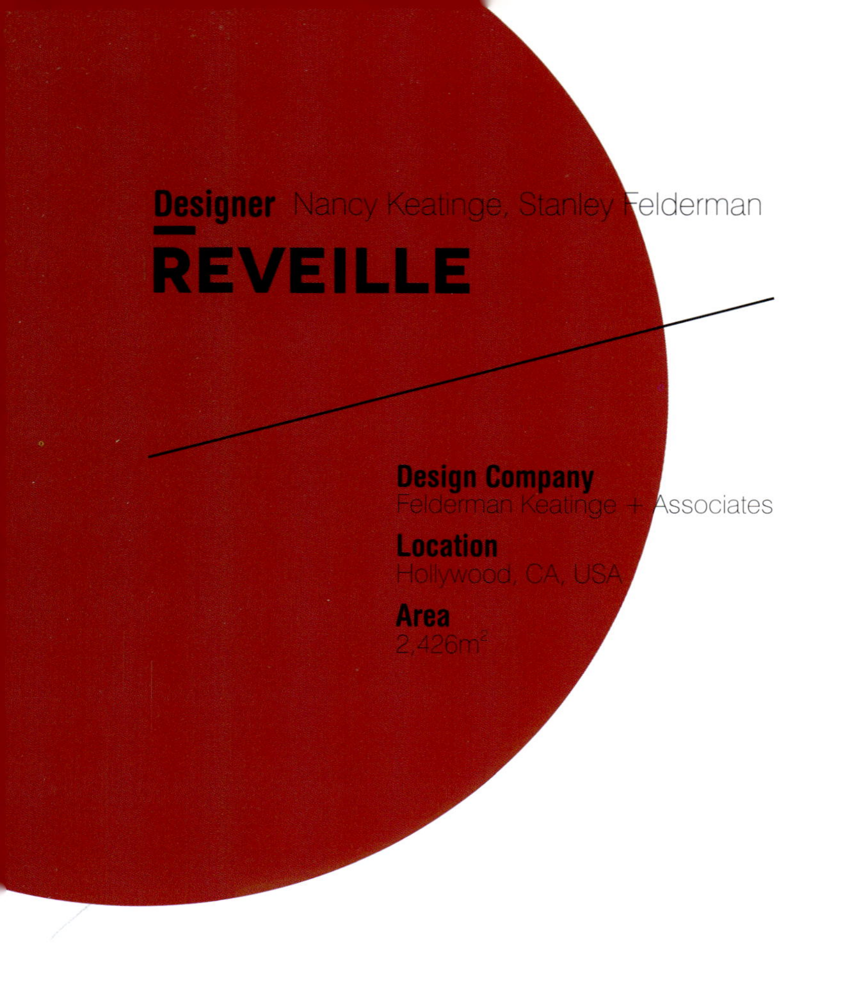

Designer Nancy Keatinge, Stanley Felderman

REVEILLE

Design Company
Felderman Keatinge + Associates

Location
Hollywood, CA, USA

Area
2,426m²

For years, client Elisabeth Murdoch's three LA-based companies were located in various locations on the Universal Studios lot. Not only did this makes it difficult to bring the whole staff together to discuss the interrelated needs of each entity, but it deprived the combined corporate body of an architectural identity, a look and a place tailored to its collective brand.

As soon as FKA and the client saw the landmark building in Hollywood, they recognized its potential. With its red brick façade, exposed structure, and steel framed windows, the building had a sense of history about it, and its undisciplined ductwork and mechanical systems gave the interior an ad hoc feel that seemed right.

The architects describe the office as "young people's space", a flexible atmosphere that allows the employees to imprint their own styles, preferences, and working patterns upon it – to lay back and put their feet up, or to pull an all-nighter when it's down to the wire. White laminate furniture over fin-ply makes for a neutral, clean palette, simplifying the space and clearing away visual clutter; ambient task-light fixtures with a retro, 50's look reflect the original building, evoking the glamour of old Hollywood while answering to the needs of the modern workspace. With the fluctuation in staff population, FKA designed custom furniture that expands and retracts. Moveable privacy panels also allow for flexible changes, without rearranging furniture. The north and south wings of the office are separated by a welcoming light-filled courtyard – formerly a motor courtyard – now closed off and replete with palm trees and plants, acting as an outdoor eating area and the main social nexus of the scheme.

FKA put the project together on a strict budget, preserving as much of the original building envelope as possible for maximum economy. The client remained very much involved throughout the process, and all three of the constituent companies offered their input at every stage. The complex program that emerged was adapted to the site with optimal efficiency for a subtle, effective space that does a lot with a little. Throughout the process FKA stayed within budget, even with a shortened timeline.

WORK
HARD
AND
BE
KIND

REVEILLE
IN PRODUCTION
THE BURIED LIFE

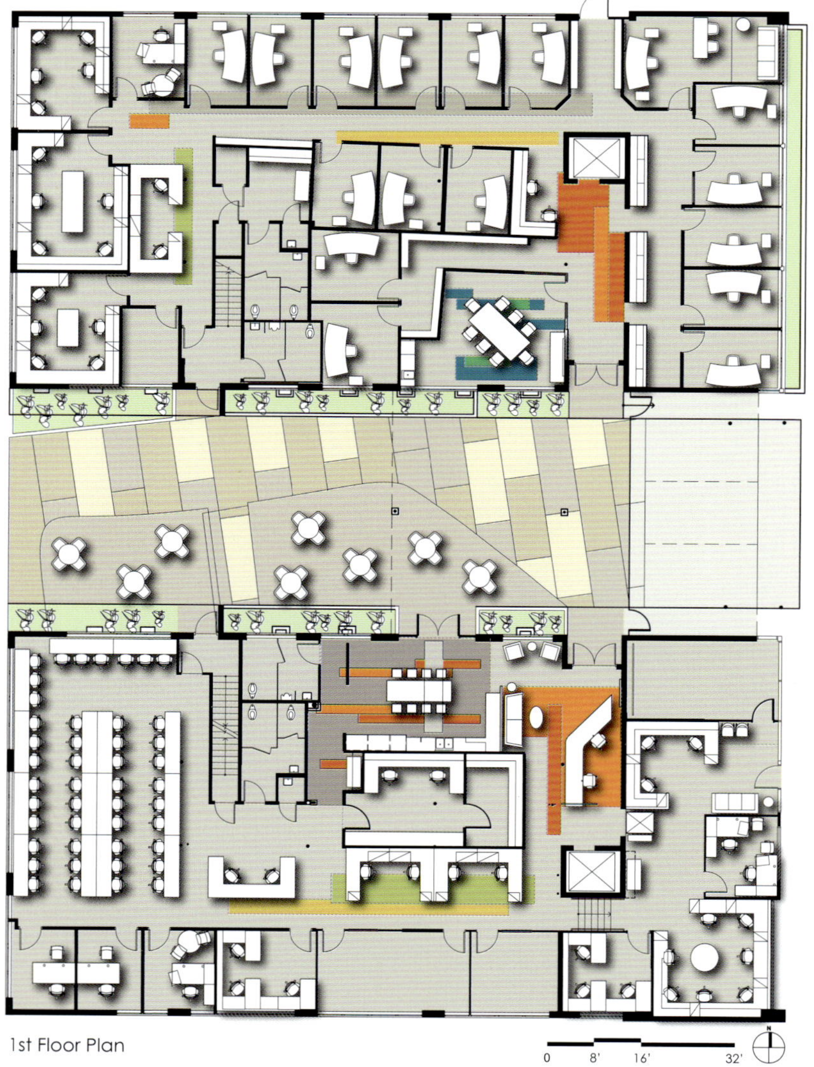

1st Floor Plan

2nd Floor Plan

多年来，客户伊丽莎白·默多克的三个洛杉矶公司都建在了环球制片厂的不同地点。这不仅使员工们聚在一起讨论相关需求变得困难，更使联合公司无法拥有一致的建筑标识、外观和地点。

FKA 公司和客户一看到好莱坞这个地标性建筑，就意识到了它的潜力。红砖外墙、裸露式结构和钢制框架的窗户为这座建筑平添了些许沧桑之感，无序的管道和机械系统使建筑内部看上去十分特别，好像就应该是这样才对。

建筑师们将办公室描述成“年轻人的天下”，这种随性的氛围能够让员工们拥有自己的风格、喜好及工作模式，比如，躺在靠背上把脚放上办公桌，或是开夜车到天亮。层板之上的白色层压材料家具形成了一个素净整洁的平面，简化了空间，扫清了视觉上的阻碍；复古式的固定工作照明环绕四周，50 盏灯同时照着原始建筑，在满足现代工作环境需求的同时也散发出了历史悠久的好莱坞的魅力。由于工作人数的可变性，FKA 公司设计定制了可展开、可收缩的家具。可移动的私人控制板也能进行灵活的改变，无需重组家具即可实现。办公室的南北两翼由充满阳光的庭院分隔开，庭院之前是停车场，现在已经关闭，并且种满了棕榈树和其他植物，作为户外餐饮区并充当了进行社交活动的场所。

FKA 公司在该项目中严格控制预算，既要尽可能地保留原建筑的外形，又要做最大限度的节省。客户在整个过程中都十分配合，三个合组的公司在每个阶段都付出了努力。这个复杂项目经过最优质的改装成为了一个设计巧妙、利用率高、花费甚少却益处颇多的空间。在整个装修的过程中，公司不仅没有超出预算，而且还缩短了工期。

BAUHAUS

ART

Designer i29 interior architects

TRIBAL DDB OFFICE

Constructor
Slavenburg

Interior Build
Zwartwoud

Area
650m^2

Material
White epoxy flooring, felt, hpl, steel

Photographer
i29 interior architects

Tribal DDB Amsterdam is a highly ranked digital marketing agency and part of DDB international, worldwide one of the largest advertising offices. i29 interior architects designed their new offices for about 80 people.
With Tribal DDB our goal was to create an environment where creative interaction is supported and to achieve as much workplaces as possible in a new structure with flexible desks and a large open space. All of this while maintaining a work environment that stimulates long office hours and concentrated work. As Tribal DDB is part of an international network a clear identity was required, which also fits the parent company DDB. The design had to reflect an identity that is friendly and playful but also professional and serious. The contradictions within these questions, asked for choices that allow great flexibility in the design.
Situated in a building where some structural parts could not be changed, it was a challenge to integrate these elements in the design and become an addition to the whole. i29 searched for solutions to various problems which could be addressed by one grand gesture. At first a material which could be an alternative to the ceiling system, but also to cover and integrate structural parts like a big round staircase was needed. Besides that, acoustics became a very important item, as the open spaces for stimulating creative interaction and optimal usage of space was required.
This led them to the use of fabrics. It is playful, and can make a powerful image on a conceptual level. It is perfect for absorbing sound and therefore it creates privacy in open spaces. And they could use it to cover scars of demolition in an effective way. There is probably no other material which can be used on floors, ceiling, walls and to create pieces of furniture and lampshades than felt. It's also durable, acoustic, fireproof and environment friendly. Which doesn't mean it was easy to make all of these items in one material!
i29 always looks for choices that answer to multiple questions at the same time. They tell a conceptual story about the company, the space and the users of the space. They deal with specific practical and functional issues and they have to have some autonomous quality as well. These "levels" are intertwined; one leads you to the other. If you see how smart it serves it purpose practically it leads you to the company. If you see the powerful image that is non-depended, it leads you to the functionality, and round it goes.

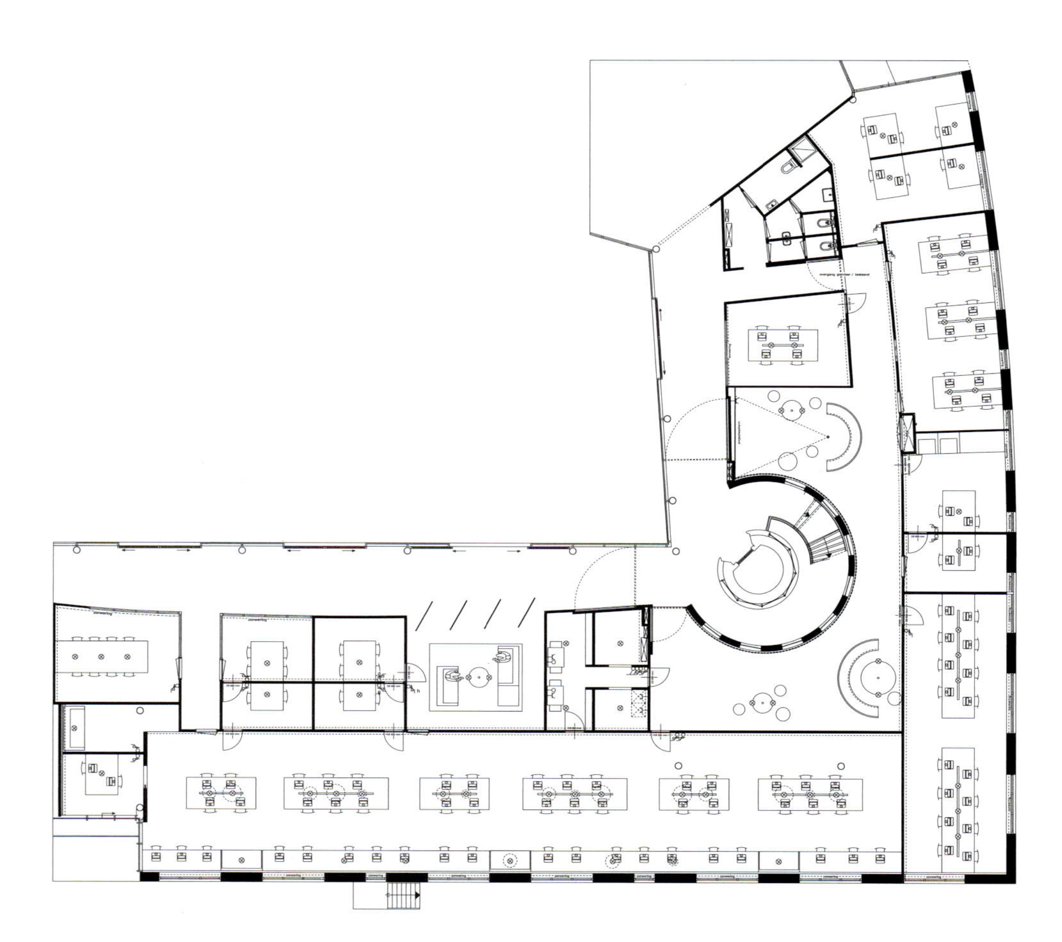

Tribal DDB阿姆斯特丹是高度知名的数字营销机构，亦是作为全球最大的广告公司之一的DDB国际集团的分支机构。i29室内设计工作室的设计师们为这个拥有80多名员工的全新办公室进行了设计规划。

这次合作的目标是为Tribal DDB阿姆斯特丹的新办公室打造一个极富创造性的互动工作环境，并在全新的空间结构中，通过灵活的书桌和大块的空地，提供尽可能多的工作区域。同时，该设计亦能保有一个工作环境，来激励员工进行长时间的专注的工作。作为DDB国际集团的一部分，Tribal DDB的新办公室需要有明确的标识，以符合母公司的统一形象。设计需要凸显DDB既亲切活泼又专业严谨的特质。因此，该设计需要有很大的灵活性。

新办公室所在的建筑中，有一部分结构无法改动。因此，如何在设计中将这些元素结合在一起，并使之成为整体的一个附加部分是设计师面临的一个挑战。i29室内设计工作室找到了一种可解决以上多种难题的方案。首先，设计师需要选择一种可替代天花板的材料，它能够像圆形的楼梯一样覆盖、整合结构性模块。此外，隔音效果也成为很重要的一项，因为它要符合激发创意且实现空间利用最优的要求。

这让设计师想到了使用织物。它不仅生动俏皮，还能在概念层次上给予强大的映像。它的吸音效果非常好，因此能在开阔的空间内开创良好的私密性。而且还可以用它有效地遮盖拆卸的痕迹。恐怕没有比毛毡更适合用在地板、天花板、墙壁、家具和灯罩上的材料了，它十分耐用，能够吸音、防火，还很环保。但这并不代表仅用一种材料去设计所有东西是件容易的事！

i29室内设计工作室总是在寻求能同时解决多种问题的方案。他们讲述了一个关于公司、空间和空间使用者的概念性故事。他们能提供特定的、可行的、功能性的设计，也拥有自己的独特之处。这些平面互相盘绕，总有一个引领着你找到另一个。当其设计引领你进入公司时，你会发现设计的实用性有多强；当你循序发现其功能目的时，你也已经看到它无支撑设计的强大形象，反之亦然。

Designer Nancy Keatinge, Stanley Felderman

HOLTHOUSE CARLIN & VANTRIGT LLP

Design Company
Felderman Keatinge + Associates

Location
Los Angeles, CA, USA

Area
2,670m^2

FKA identified a couple of "cues" for their expanded design strategy. The client's initial attraction to this particular locale had been the views; it was also a big enough interior to bring together two working groups within the company that had formerly been housed in different locations.

The goal, from FKA's perspective, was to create a sense of community, and to let everyone get a chance to share the views and the abundant daylight – and to do all this within a fairly strict budget. Pursuing the theme of community, FKA designed the training facility, conference room, lunchrooms, and gathering spaces so that they connect fluidly with the rest of the building on the east and west sides. Visual privacy was not essential, so the designers could deploy large expanses of glass and expose to open up those views. But acoustic privacy was a key, so FKA had to use laminated glass and make other provisions in order to dampen sound.

The clients had originally wanted to use existing furniture, but FKA convinced them that it would actually be less cost efficient in the long run to do so. For other fixtures, FKA opted for organic textures, stone and wood flooring, and carpeting with a natural feel, to invigorate employees working long hours. FKA turned to the team at Haworth for the workspace furniture; the manufacturer took a hands-on approach in co-creating a layout for the workstations unique to the project, which required maximum functionality in the minimum amount of space. The clients also liked the different finishes and materials available from Haworth. (They were able to get higher-end materials at a lower price point.)

Adding even more visual interest, FKA was commissioned to create original artwork to enhance the interior: firm principal Stanley Felderman installed a 20-meter mural and other pieces, and the conference room tables were an original FKA design.

Response from the employees has exceeded expectations, and the client learned just how much design can do. They say their customers look at their company in a whole new way, and that business and employee morale have both taken a big jump.

NOT.AFRAID
RUBELL FAMILY COLLECTION
arthur dove: a retrospective

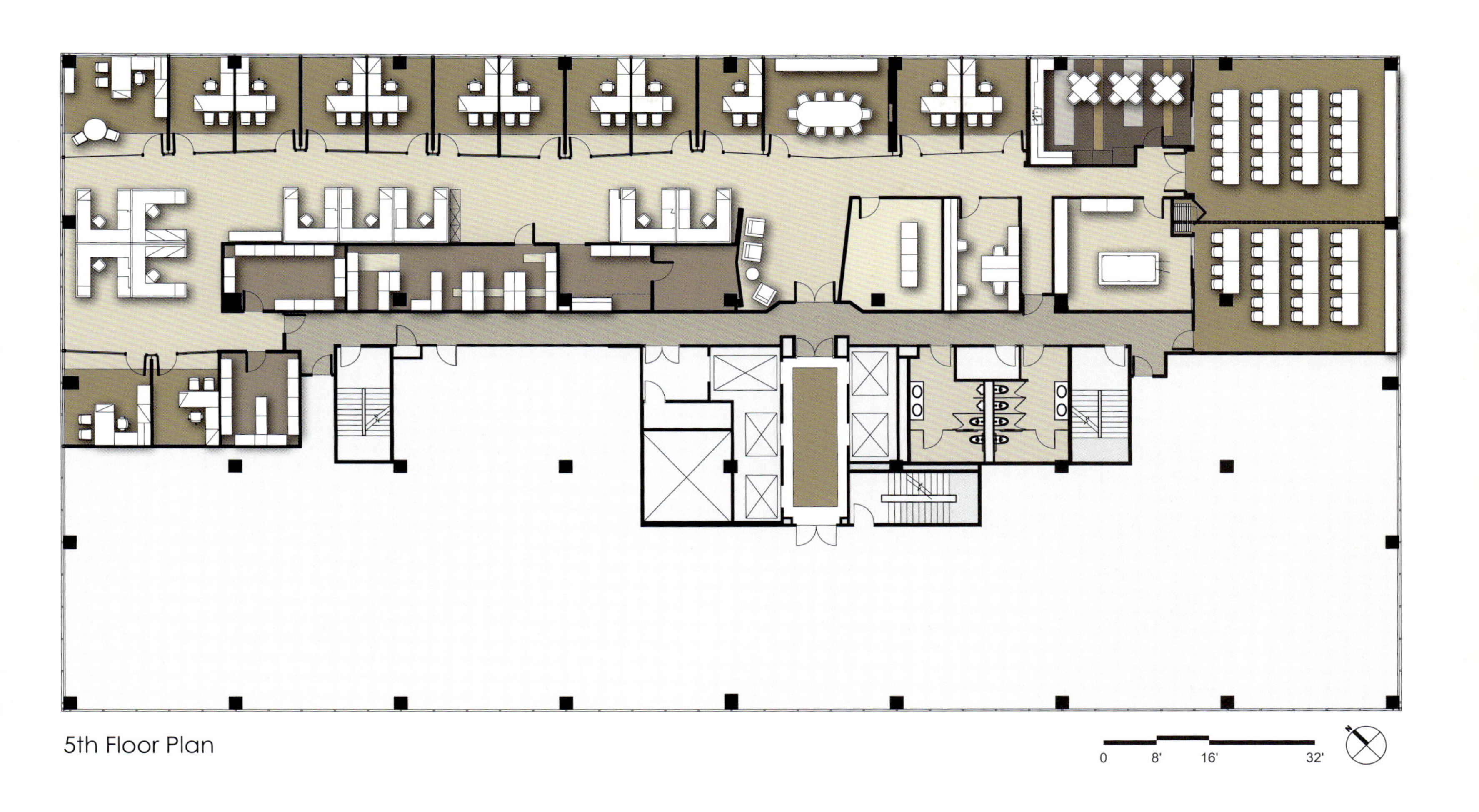

5th Floor Plan

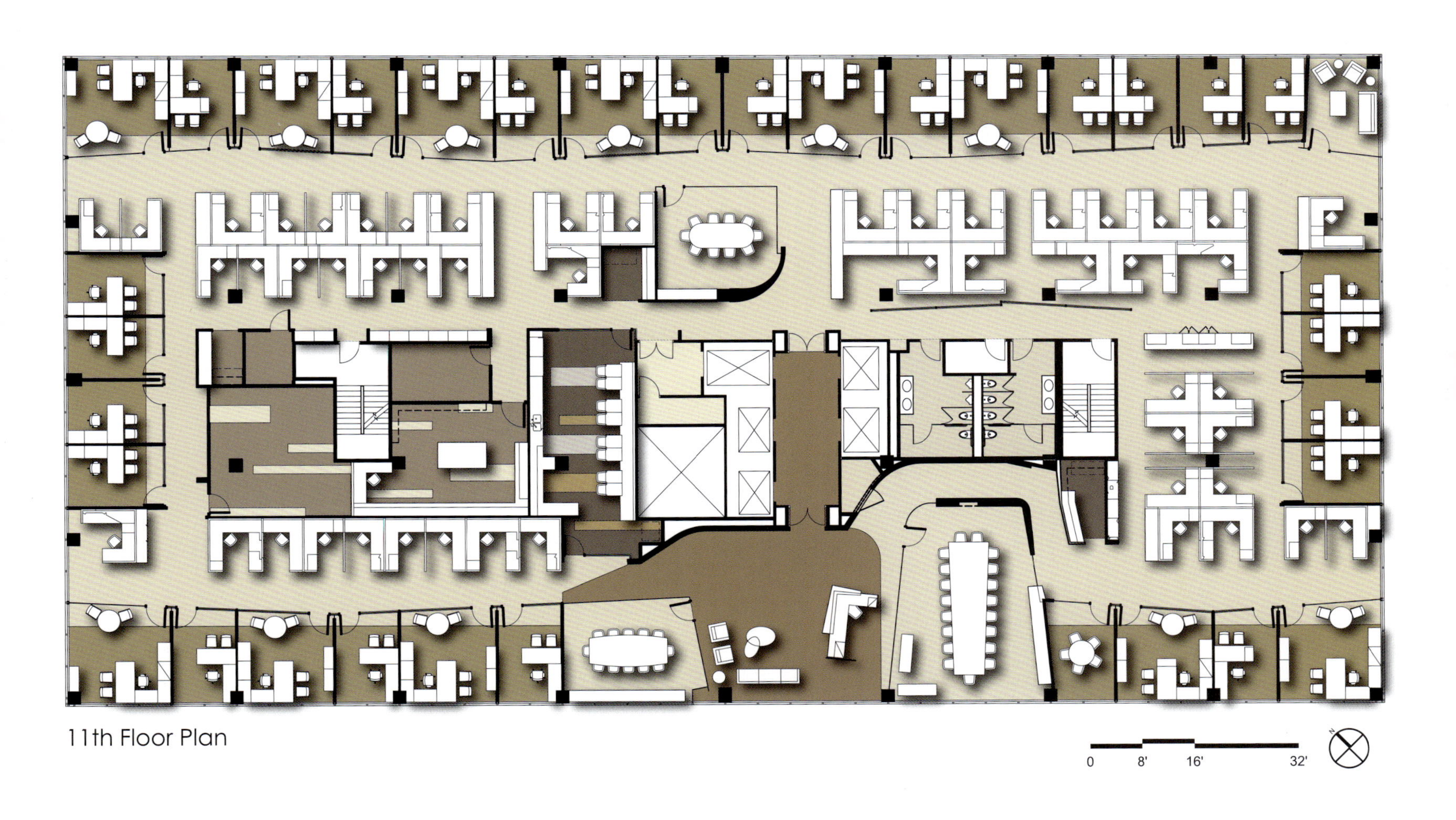

11th Floor Plan

为了拓展设计战略，FKA 公司提出了几个提议。客户最初对这个地方产生兴趣是因为这里的外部景色，而且其室内空间宽敞，可容纳两个工作组共同工作，而不必像以前那样分设两地。

从 FKA 公司的角度来讲，该设计的目标是营造一种社区感，让每一个人都有机会分享不同的观点和充足的阳光——而这一切都是在严格的预算范围内完成的。为了追求这种社区感，FKA 公司设计的培训设施、会议室、食堂和聚会场所都与建筑内部东、西两个方向搭配得当。视觉隐私并不是主要的，所以设计师采用大块的玻璃，营造更多的视觉空间；但是听觉隐私很重要，因此 FKA 公司设计师采用夹层玻璃来达到隔音的效果。

客户最初想用现有的家具以节省成本，但是 FKA 公司却不建议这么做，因为从长远考虑，这样做实际上事与愿违。在其他固定设施的选择上，FKA 公司采用有机纹理的石头和木质地板，地毯质地柔软，即使员工们长时间工作也不会觉得很累。至于工作区的家具，FKA 求助于哈沃斯的团队，与那里的制造商一同设计并完成了工作区独特的布局。该布局仅占用了最小的空间，却实现了功能最大化。客户也很喜欢从哈沃斯采购的各种装饰和材料，可谓物美价廉。

为了添加更多有趣的视觉效果，FKA 公司还受托设计了一些原创艺术作品，以增强室内设计感：公司内一幅高约 20 米的壁画和一些其他的作品都由主设计师斯坦利・费尔得曼亲自安装。会议室的桌子也是 FKA 公司的原创设计。

员工的反响比预期还好，客户也因此明白设计二字不仅仅是简单的装潢。客户说，他们的客人对公司的焕然一新刮目相看，而且公司的业务量和员工士气也大大提高了。

Designer Taylor Smyth Architects

HUDBAY MINERALS INC. OFFICES

Location
25 York Street, Toronto, Canada

Area
2,415m²

Photographer
Ben Rahn/A-Frame

The new head offices for HudBay Minerals Inc. occupy the 8th floor of the Telus Building at 25 York Street. The client is a Canadian integrated mining company principally focused on the discovery, production and marketing of base metals.

The interiors have been designed to LEED Silver CI, in keeping with the base building LEED standards, and to reinforce the company's environmental commitment to sustainable strategies.

The design was inspired by images from the company's extensive photo library, ranging from the mining of the raw materials to the smelting processes and the finished products.

The story begins at the elevator foyer, where actual ingots of raw zinc, containing the HudBay stamp, have been mounted along each wall. The ceiling is composed of panels of recycled spun aluminum, whose open cells allow light from above to filter through. This also evokes the netting that wraps the ceilings of the mines.

Stepping into the reception area, a dramatic view of the Toronto skyline is framed on one side by three freestanding stone walls that define a large meeting room and on the other by a commissioned installation by Toronto artist Dennis Lin. The walls are comprised of rough split face stone, inspired by a HudBay photo of vertical cliffs that have been hewn from the rock, while the installation of copper and walnut evokes both the strata of the earth and the sinuous line of copper as it is poured into moulds in the company's foundries.

HUDBAY
MINERALS
HUDBAY
MINERALS

HudBay 矿业有限公司的新总部设在约克街 25 号，Telus 大楼的八层。HudBay 是一家加拿大综合采矿公司，主营业务为普通金属的开采、生产和营销。

HudBay 矿业有限公司的内部设计依照绿色能源与环境设计银奖标准（LEED Silver CI），既符合整体建筑的标准，又强调了公司环境的可持续发展战略。

本案的设计灵感来自公司照片库中覆盖内容甚广的图像，从原料的开采、冶炼过程到成品。

让我们从前厅的电梯说起，四壁挂满带有 HudBay 印记的锌锭。天花板由回收的铝板制成，阳光可以从这种开放式结构中穿过。这也让人想起了包裹矿井顶棚的网。

走进接待处，映入眼帘的是多伦多市的全景，令人赞叹不已。一侧是由三面独立的石头墙围成的大型会议室。石头墙由岩石上凿下的粗糙碎石构成，其灵感来自一张 HudBay 悬崖峭壁的照片。另一侧由多伦多艺术家丹尼斯·林受邀设计，墙壁上装饰有铜和胡桃木，会让人想到地层及将铜液注入公司铸造用的模子里时产生的复杂纹路。

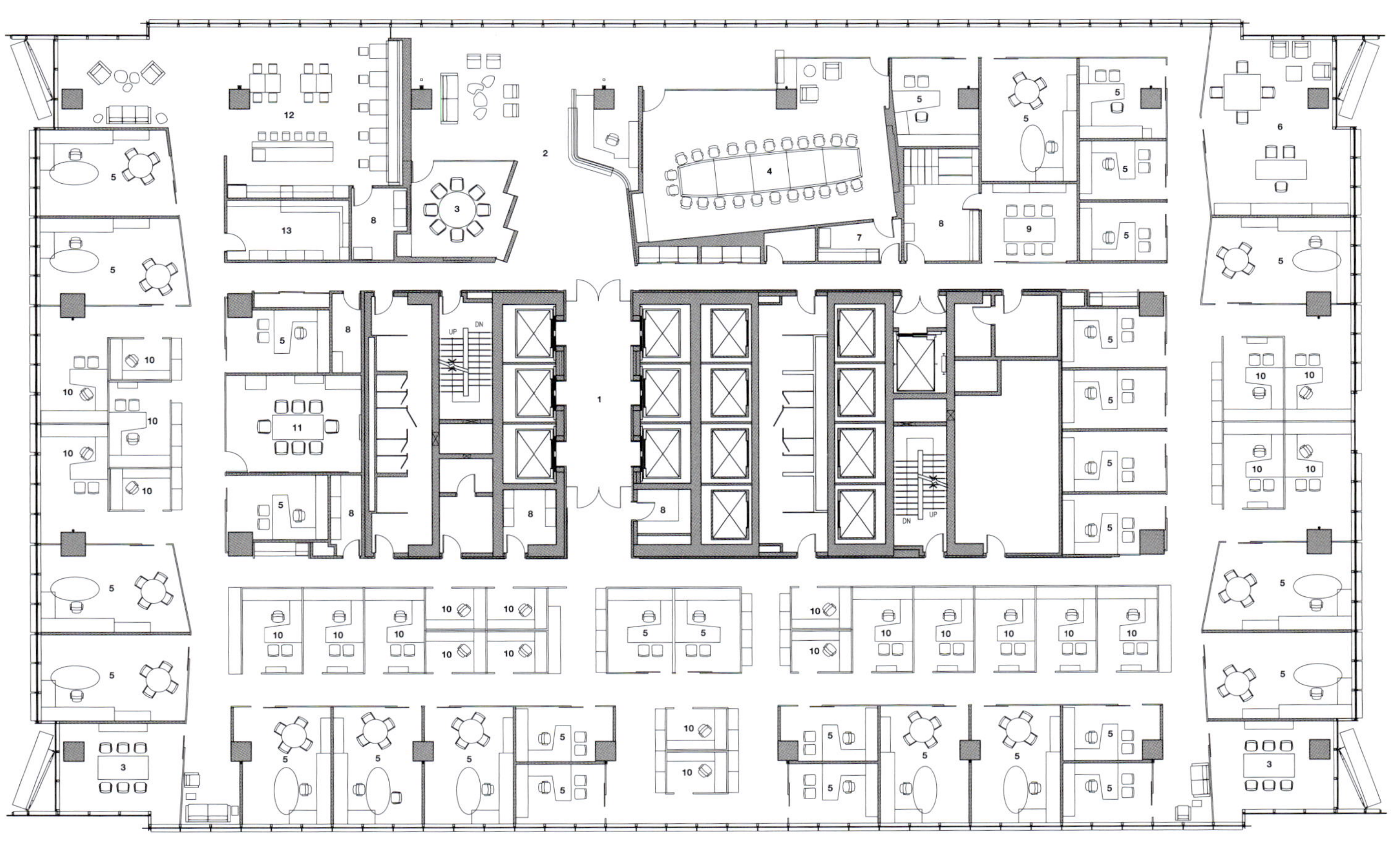

1 elevator lobby
2 reception
3 meeting room
4 board room
5 office
6 ceo office
7 servery
8 storage
9 legal library
10 workstation
11 map room
12 lunch room
13 copy room

HudBay Minerals Offices
FLOOR PLAN

Dave Garofalo

Designer Nancy Keatinge, Stanley Felderman

ICRETE

Design Company
Felderman Keatinge + Associates

Location
Beverly Hills, CA, USA

Area
855m^2

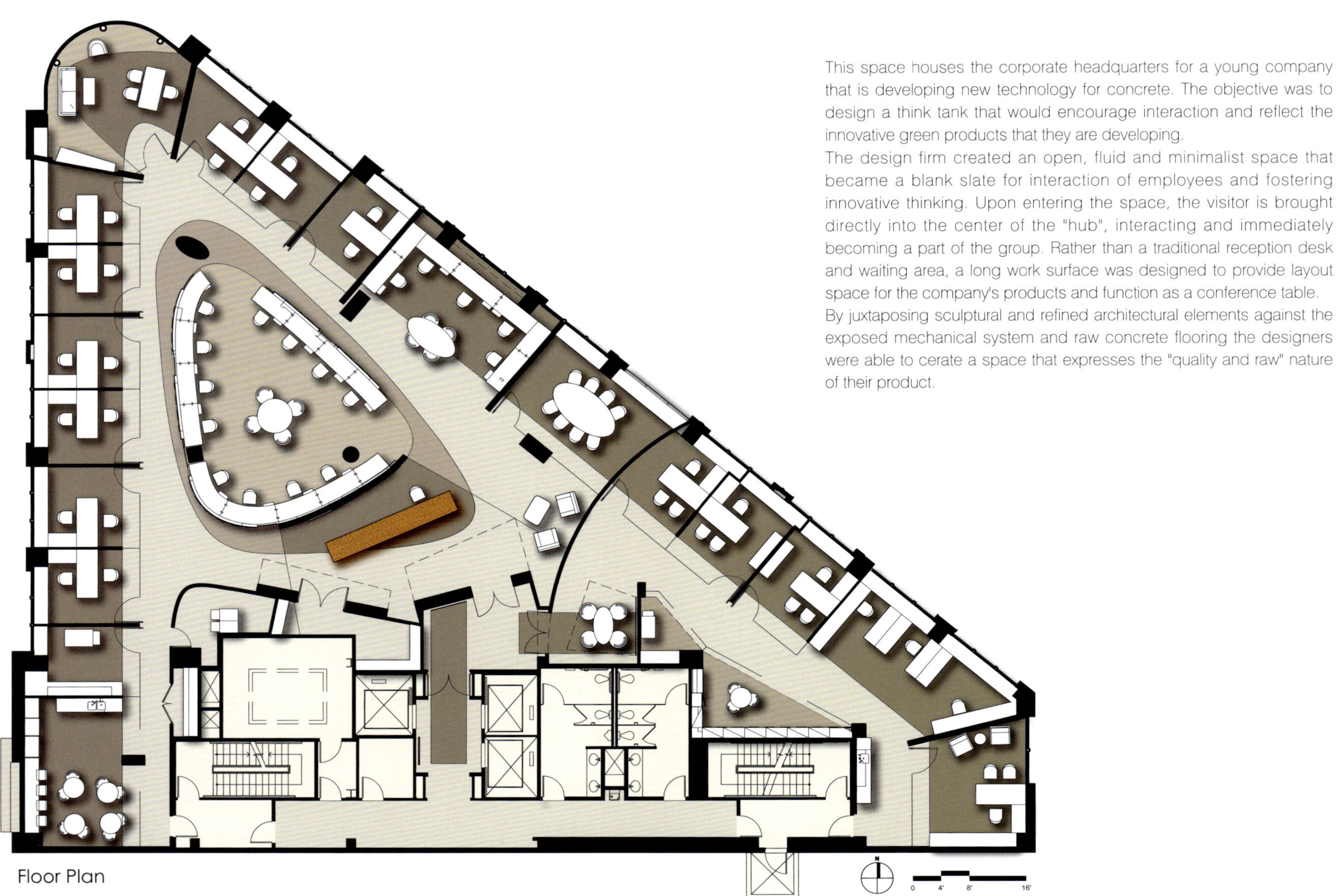

Floor Plan

This space houses the corporate headquarters for a young company that is developing new technology for concrete. The objective was to design a think tank that would encourage interaction and reflect the innovative green products that they are developing.

The design firm created an open, fluid and minimalist space that became a blank slate for interaction of employees and fostering innovative thinking. Upon entering the space, the visitor is brought directly into the center of the "hub", interacting and immediately becoming a part of the group. Rather than a traditional reception desk and waiting area, a long work surface was designed to provide layout space for the company's products and function as a conference table.

By juxtaposing sculptural and refined architectural elements against the exposed mechanical system and raw concrete flooring the designers were able to cerate a space that expresses the "quality and raw" nature of their product.

该办公空间是一家新兴公司的总部，这家公司主要从事新型混凝土技术的开发。本案设计的目的在于创造一个智慧中心，既可以促进交流又可以体现出他们正在开发的绿色产品。

设计公司创造了一个极简抽象的空间，开阔而流动，如空白的石板一样可以促进员工之间的交流，培养员工的创新性思维。一进入这个空间，来访者就会直接进入中心区，立刻融入其中，成为其中的一员。接待台和等待区域也有所创新，长长的工作台可为公司产品提供展示空间，也可以充当会议桌。

设计师将精美的雕刻和建筑元素与显露在外的机械结构和原始混凝土地板并置，形成鲜明的对比，以此营造一个可以展现公司产品的“质感与原始”特质的独特空间。

Designer Jose Abeijón Vela, Miguel Fernández Carreiras

MAXAN'S OFFICE

Design Company
a.f. abeijón-fernandez arquitectos

Location
C/ Comandante Barja 9, A Coruña, Spain

Area
632.46m^2

Since MAXAN is an advertising agency, this project relates working places and the job done inside them. The designers apply the concept of image, technique interpretation and marketing and advertising tendencies to the physical elements that make up the space.

As they couldn't modify floor, the designers try to emphasize the height of ceilings. White color gives walls a dramatic quality and colored areas forms only on certain elements. For these areas they have used the colors that appear in MAXAN's logotype.

The designers relate both open areas and exchange of ideas and both visual permeability and creativity. All these concepts promote a modern image.

In shorts they try to create a small temple for the research and innovation in graphic arts applied to advertising.

QUEREMOS QUE ENAMORES.
LA GENTE TE MIRE Y TE DESEE
SE DETENGA CUANDO APAREZCAS.
GUAPO, INTELIGENTE E
PUEDES TENER UN CARISMA
SI NO ENAMORAS, NO EXISTES.

QUEREMOS QUE
LA GENTE TE MIRE Y
SE DETENGA CUANDO
GUAPO, INTELIGENTE
PUEDES TENER UN
SI NO ENAMORAS

MAXAN
REFLEXIONES
PERFIL
EQUIPO
CLIENTES
cross section
section

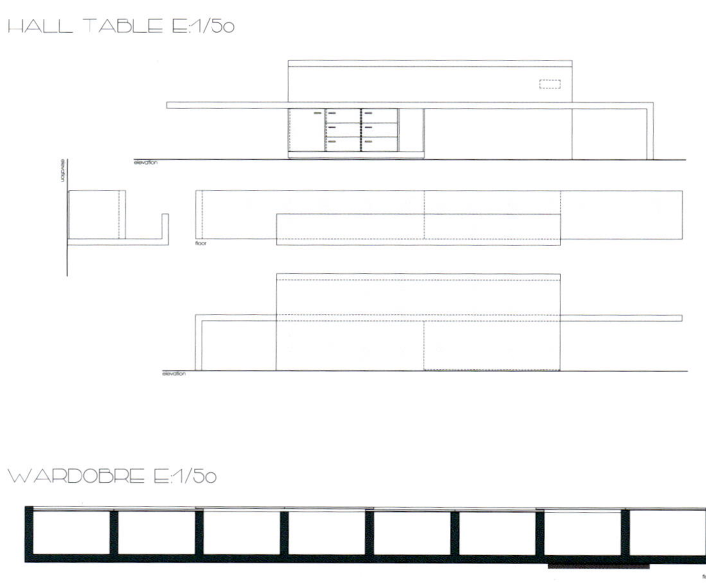
HALL TABLE E:1/5o
WARDOBRE E:1/5o

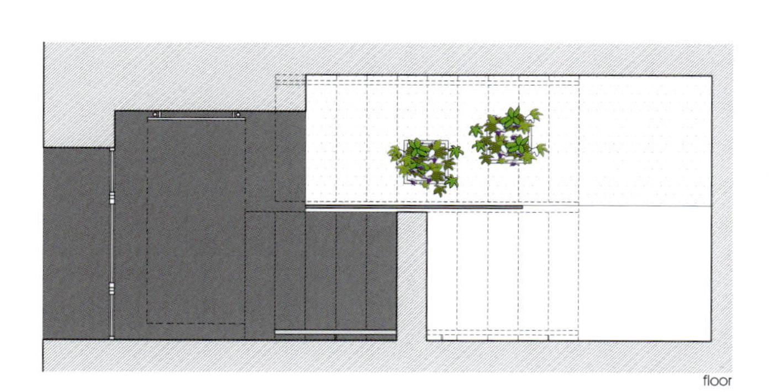
floor

E:1/5o

MAXAN 是一家广告代理机构，因此，本案将涉及办公场所和场所里的工作。设计师将图像和技术分析的理念、市场营销和广告趋势应用到有形的元素中来打造这一办公空间。

因为地板无法改动，设计师便尽力突出天花板的高度，将墙壁刷成白色以产生强烈的视觉冲击，并在一些特定地方采用彩色的空间设计。这些区域所用到的颜色都是从 MAXAN 公司的商标上提取的。

设计师不仅设计了开放空间以方便员工交换想法，还将视觉渗透的概念和创造性融合到设计之中。所有这些设计理念都使 MAXAN 办公室具有一种现代的形象。

总而言之，设计师尽力创造出一个广告图形艺术的研发和创新空间。

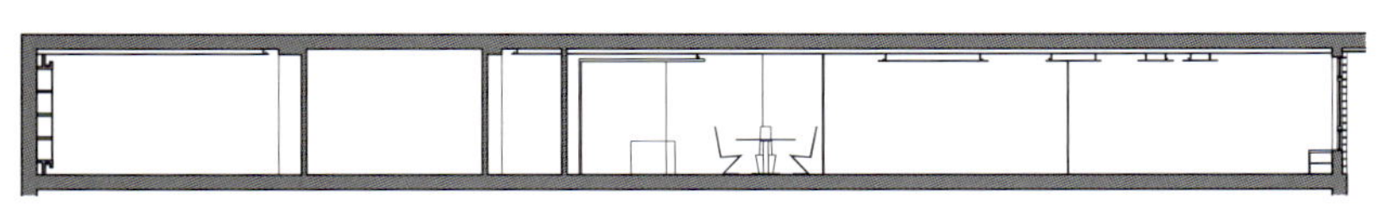

SECTION
FLOOR E:1/150

MAXAN

Designer Kapil Aggarwal, Nikhil Kant

NOCC OFFICE

Design Company
Spaces Architects

Location
223, Udyog Vihar, gurgaon, Haryana, India

Area
929m^2

Photographer
Akhil Bakshi

The concept has been derived using abstract shapes, panels, forms and the use of pastel shades. It's a spatial experience which one discovers as walking through the corridor. An attempt has been made to create multiple views beyond partitions and connecting the whole area to give an open office environment.

The office interior of 929 square meters has a main studio of 30 seats with multiple NOCC rooms. Two office areas which are supporting NOCCs are placed on either side. The main studio has ascending stepped platform in curved form reflected on the ceiling. The glasses of the open studio are inclined in abstract shape placed along the side. The abstract curved partition at front defines physically the space and also connects visually to the office space. The geometric and abstract forms with the use of pastel color define the space. Egg shaped meeting room intersects with rectangular meeting room placed at 45 degree angle. The elliptical meeting room has a bright green pastel shade contrasting with dark teak cladding on the square meeting room. The front of the rectangular meeting room has been laminated with mirror effect laminate with circles having transparent glasses. This creates illusion when one proceeds from the entrance towards the meeting room. The ceiling outside the rectangular meeting room is projected outside with backlit acrylic sheet. The same impression is given on floors by using dark colored tiles which gives the feel of shadow of the ceiling. This dark colored tile is also used to create transitional spaces throughout.

The interior space has been perceived as an art form with free flowing spaces, elements, their relationships and response to humans. A frame with elements forms a different canvas viewed from different directions. It's an experiment to connect floor, wall and ceiling. The studio furniture has all the tables in curved form. Each table has different cladding with multiple sizes of different shades of mica highlighted by cove lighting.

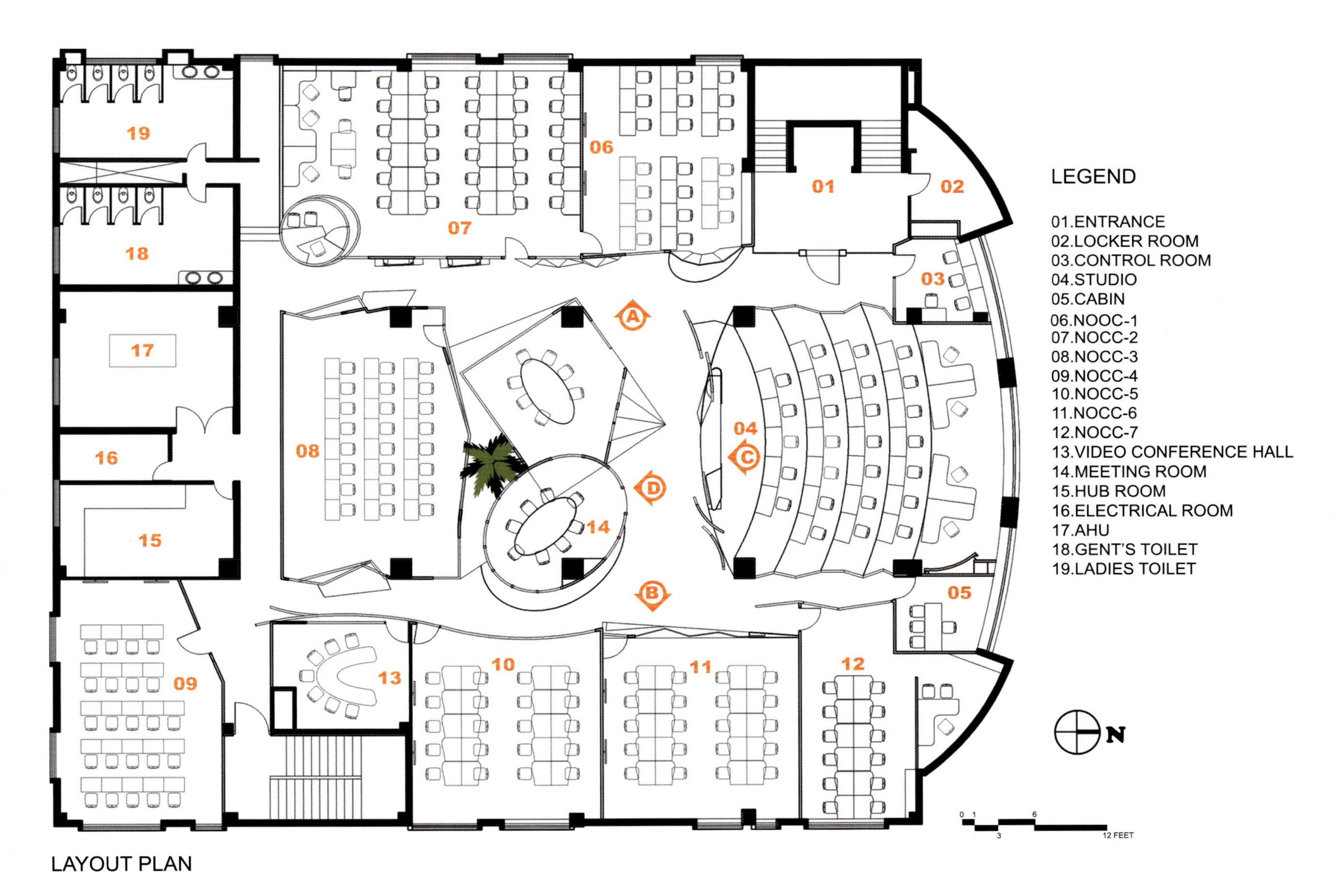

LAYOUT PLAN

本案的设计理念通过使用抽象的形状、嵌板、样式和柔和的色调展现出来。走在走廊上，你就会发现这是一次空间体验。设计师尝试跨越分割，创建多角度的视野，将整个区域相连以营造一个开放的办公环境。

办公区室内面积为929平方米，包括一间可容纳30人的主工作室和其他NOCC办公室房间。左右两侧各设一个辅助办公区域。主工作室设有弯曲的阶梯状平台，与天花板的设计相呼应。开放式工作室的玻璃以抽象的形状沿着一边倾斜下来。前方抽象而弯曲的隔墙将空间进行有形分割，并在视觉上与办公区相连。柔和色调的抽象几何形状勾勒出整个空间。鸡蛋形状的会议室与长方形会议室交叉成45度角。椭圆形会议室的明绿色调与深色柚木包裹的方形会议室形成鲜明的对比。长方形会议室的前面制成透明玻璃圆圈，层压成镜面效果。这会使人从入口走到会议室时产生错觉。长方形会议室外的天花板上投射有背光式的亚克力板。地板采用了深色瓷砖，同样使人形成错觉，好像它是天花板的影子。这种深色的瓷砖同样用于打造整个过渡空间。

自由流动的空间、元素以及它们之间的关系和带给人们的感受，使整个室内空间更像一件艺术作品。从不同的方向看去，多种元素组成的框架仿佛不同的油画。这是一次将地板、墙面和天花板连为一体的尝试。工作室的所有桌子都是弯曲的形状。桌子的表面是尺寸不同、明暗各异的云母，并使用凹圆形天棚照明。

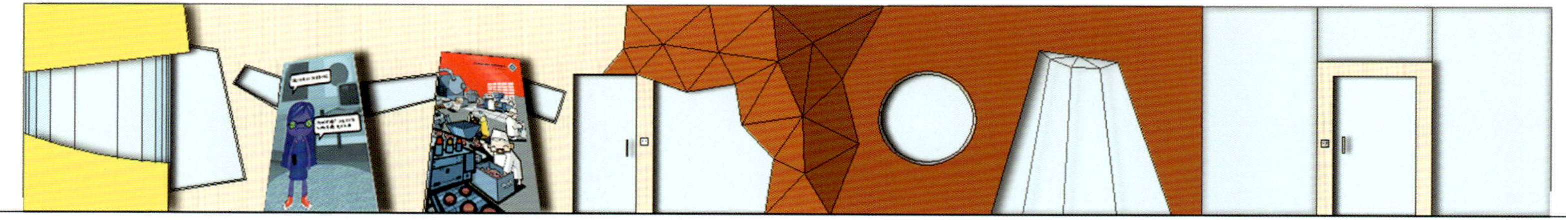

CORRIDOR ELEVATION (A)

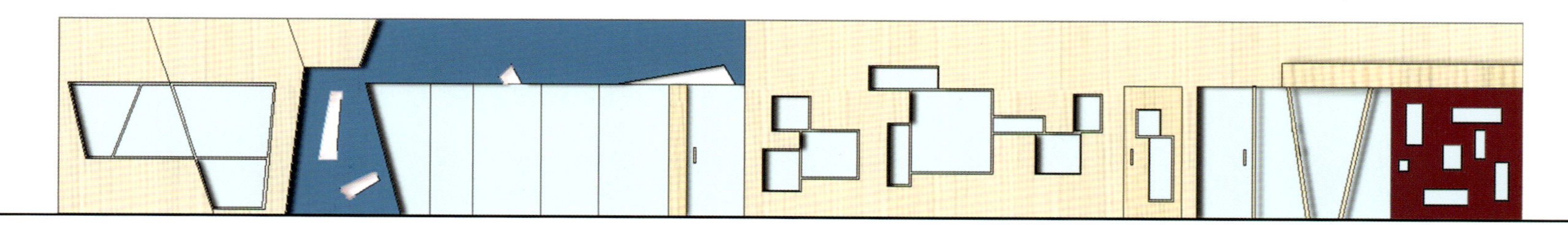

CORRIDOR ELEVATION (B)

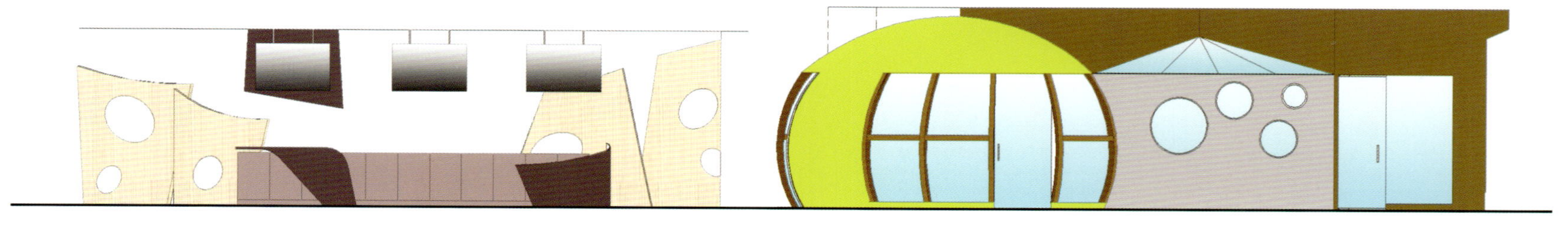

MAIN STUDIO ELEVATION (C)

MEETING ROOM ELEVATION (D)

EarthLink

AXA Tech -2

Designer Mr. Dipen Gada, Mr. Yatin Kavaiya, Jiten Tosar

OFFICE WRITE-UP

Design Company
Dipen Gada & Associates

Landscape Designer
Mr. Tulsibhai Narola

Text
Shalini Pereira

Area
$93m^2$

Photographer
Tejas Shah

Concept — back to basics

Creating a space that is highly functional as well as minimalistic and contemporary through the exploration of raw and basic building materials.

Trying to create a truly green office, which requires minimum energy consumption and with maximum efficiency of the space.

Converting what was basically a two-bedroom apartment into an open plan designer's office.

Green design

There was also a deliberate step towards making this office more energy efficient and green by the use of cool coat paints, CFL lamps, reusing the flooring of the old office and using materials in their basic form.

Use of direct exposed slab casting as a design element through which the use of false ceiling has been avoided. The use of basic materials has been emphasized so that major artificial treatments could be avoided. False ceiling is made only in 20% area to incorporate the A/C and lighting.

A terrace garden has been incorporated to ensure that heat insulation is provided to the studio below. This in turn also minimizes A/C consumption.

Windows have been placed in such a way as to ensure sufficient day lighting while minimizing heat penetration from outside.

Use of UPVC windows with double-glazed reflective glasses, which is having 85% heat, UV and IR rays reflectivity.

Recycled original flooring of the office.

Use of reflective curtains with their reflective specifications.

Minimum use of wood, no use of any veneer or laminate to avoid superficial look and to reduce the overall cost.

Old granite work tops from the old office have been reused.

Use of apple comps and all other flat screen which saves lots of energy.

Use of CFL lights and electronic ballast to reduce energy consumption.

Use of Diakin and Mitsubishi A/C as power efficient A/C's.

Use of highly reflective white paints on exterior walls. These paints are having neon technology reflective of UV and IR rays.

36cm thick exterior walls on south and west directions so as to achieve minimum heat transaction.

设计理念——返璞归真

设计的理念是通过对原材料和基本建筑材料的探究，创造一个功能实用、简约而现代化的空间。

尝试创造一间真正的生态办公室，使用最少的能源的同时使空间得到最大的利用。

将一个带有两间卧室的公寓改造成一间开放的设计师办公室。

生态设计

为了使办公室更节能、环保，设计师慎重地选择了清爽的墙面漆和节能灯管，重新利用老办公室的地板，尽可能减少对装修原材料的再加工。

通过使用裸露的铸造板作为设计的一个元素，避免了假平顶的使用。同时，强调使用原材料以避免大量的人工处理。假平顶只占了空间的 20%，用来遮盖空调和灯的电线。

设计师整合了露台花园以确保下面工作室的隔热效果，也使空调的损耗降到最低。

窗户的结构设计可以确保足够的自然采光，同时还能尽量减少外部热量的渗透。

采用聚乙烯双层玻璃，它可以反射 85% 的热量以及紫外线和红外射线。

二次使用原有的地板。

充分利用反光窗帘以及它的反光范围。

将木头的使用量减至最小，不使用漆合板或木地板，避免徒有虚表的外观，同时减少开支。

重复利用原来的大理石。

采用苹果公司的产品和其他平板屏可以节省大量能源。

采用节能灯和电子镇流器可以减少能量消耗。

采用日本 Diakin 压力泵和三菱空调以节约能耗。

外墙采用具有高反射效果的白色涂料漆，该涂料应用氖技术，能有效反射紫外线和红外射线。

南面和西面的外墙厚达 36 厘米，这样可以达到最小的热交换。

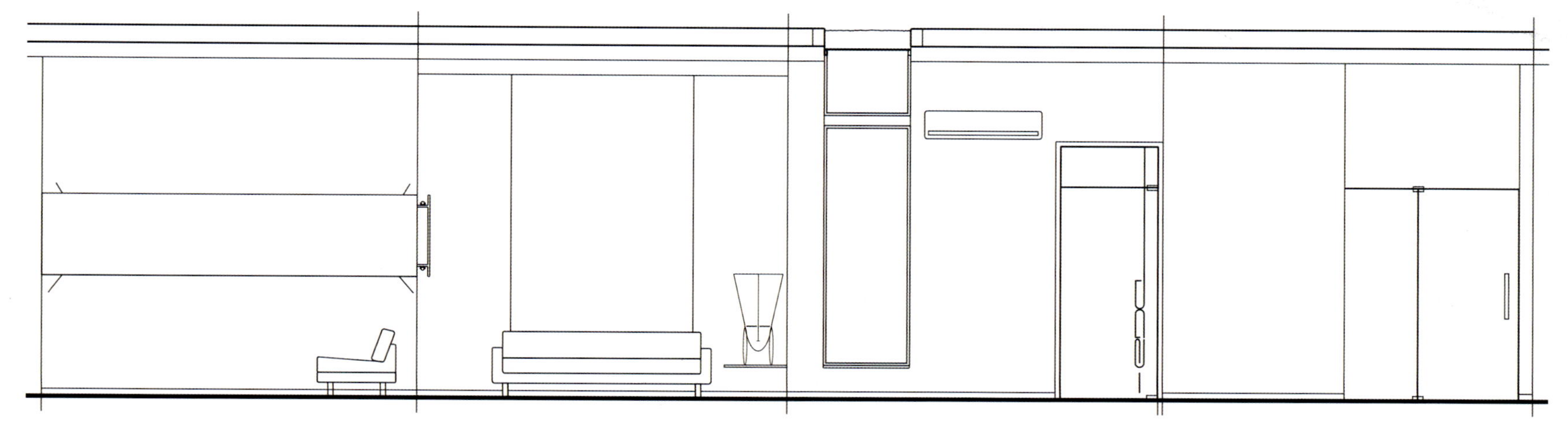

ALL WALL ELEVATIONS OF LOUNGE

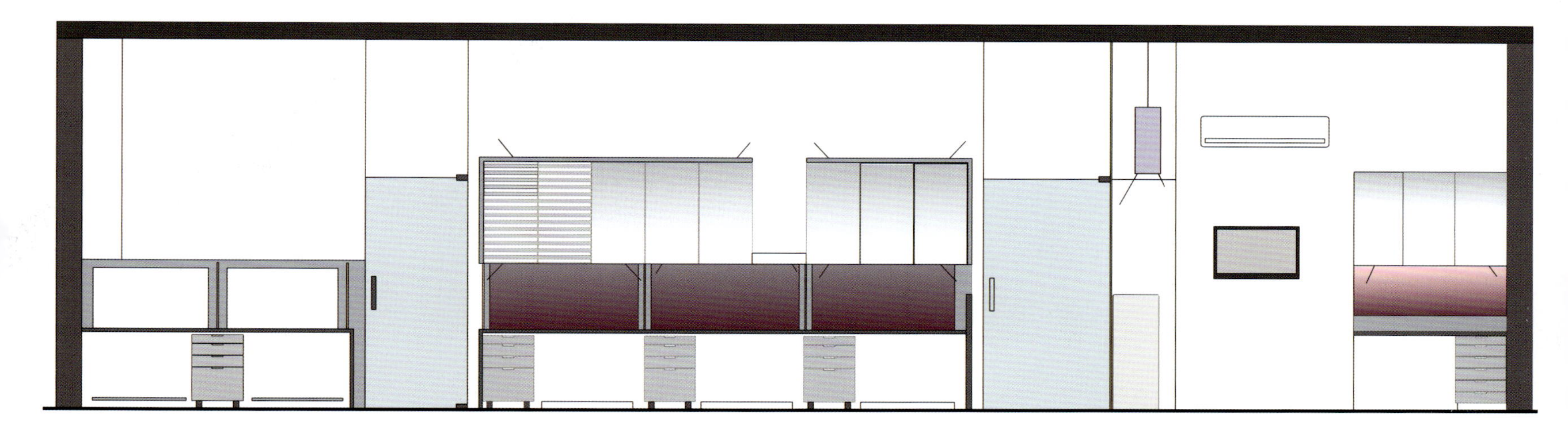

SECTION THROUGH MAIN STUDIO

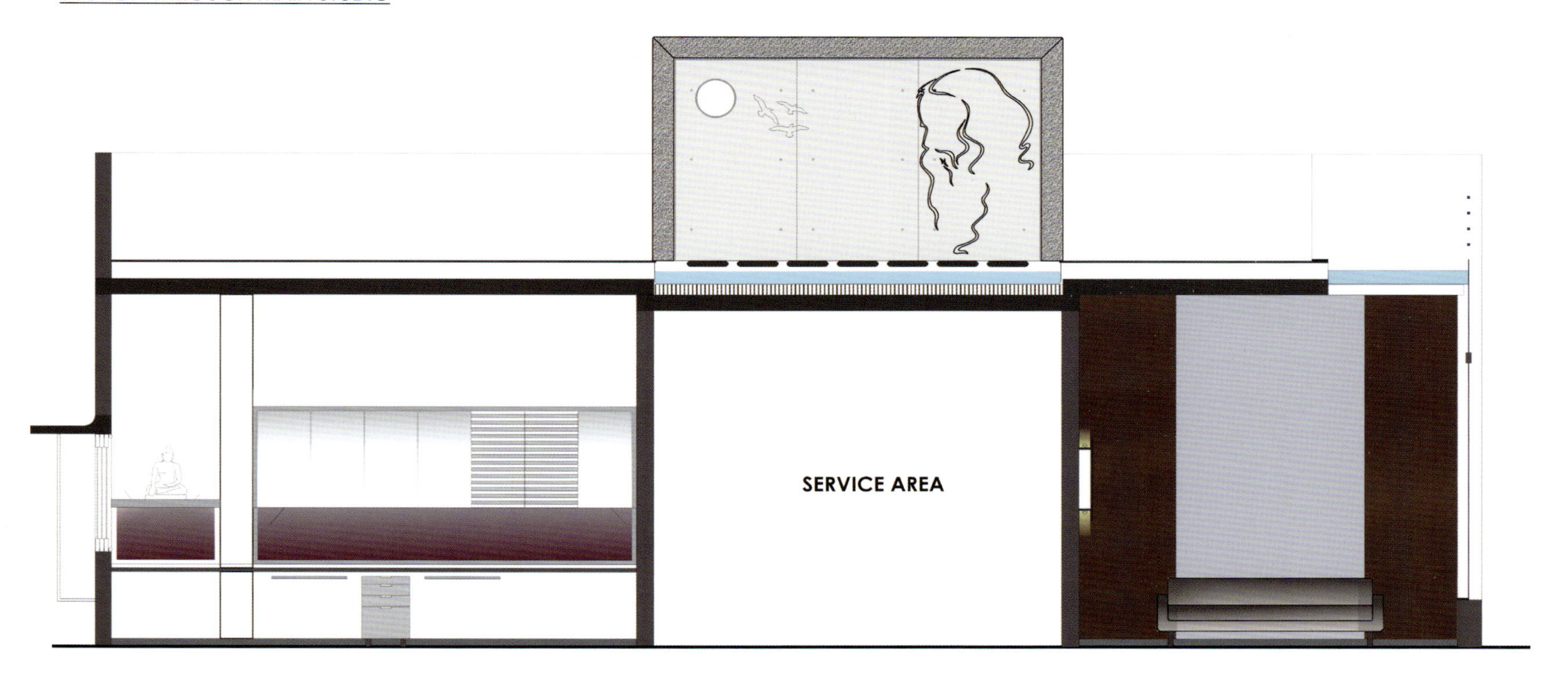

SECTION THROUGH TERRACE WATER BODY

OFFICE LEVEL PLAN

TERRACE LEVEL PLAN

Designer Nancy Keatinge, Stanley Felderman

FKA OFFICES

Design Company
Felderman Keatinge + Associates

Location
Los Angeles, CA, USA

Area
297m^2

FKA's original Los Angeles headquarters had been a high-ceilinged warehouse in Santa Monica, a space whose open floors allowed the staff to mingle easily and whose sliding garage door gave them the freedom to work and eat outside. The firm's new home, however, would be a little different, and a lot more glamorous: a class-A high-rise on the Avenue of the Stars in upscale Century City. Looking to make the space its own, the firm transformed it into an interactive environment with expansive views, giving it a casual studio-loft feeling that reflected founding principal Stanley Felderman's urban roots in New York.

The new office's classic studio layout facilitates staff interaction, with long tables that foster creative collaboration during the design process. Felderman's personal office has no door, so that people know he's always accessible. Nancy Keatinge's office has a single sliding glass door. The raised conference room floor – surfaced in elegant walnut – goes right up to the window mullions, affording unobstructed sightlines to the city below, while over the conference table hangs an Ingo Maurer Zettle 5 chandelier along with vellum prints and sketches by Felderman himself.

Other artwork by Felderman enlivens the walls throughout the office, creating an engaging, colorful atmosphere that's further accentuated by pigmented concrete floors. An extra layer of gypsum board on the building columns helps masks them, making the space feel even more airy and fluid, and the materials and text libraries are out in the open, encouraging the staff to use them and keep them up to date.

Altogether, the office imports much of the playfulness and ease of FKA's old stomping grounds, while adding a new sense of cool and sophistication. Like all of FKA's work, the firm's own office is a unique environment that's geared to reflect who its occupants are, not just where they happen to be.

FKA公司最初的洛杉矶总部是一间位于圣莫妮卡的高顶棚仓库，内部地面开阔，员工可以自由走动，还有一个滑动车库门，员工们既可以在室外工作，也可以在室外就餐。而FKA公司的新总部是一栋坐落于高档世纪城的星光大道上的甲级大厦，因稍显不同而更加迷人。为了彰显其空间特点，公司将其设计为互动式环境，视野开阔，工作室—阁楼的随意感则反映了公司创始人斯坦利·费尔得曼在纽约的城市之根。

新总部里经典的工作室布局有助于员工之间的互动，长长的办公桌能在设计过程中培养员工们进行创造性合作。费尔得曼的私人办公室没有门，员工们能随时找到他。南希·基汀格的办公室有一道单侧玻璃推拉门。会议室里一直延伸到窗框的凸起的地面铺着典雅的桃木地板，从窗台可俯瞰下面的城市。会议桌的上方悬挂了一盏Ingo Maurer Zettle5的吊灯以及由费尔得曼亲自完成的牛皮纸版画和素描。

办公室四周的墙壁上布满了费尔得曼的其他艺术作品，加上涂色的水泥地板营造出迷人而多彩的氛围。办公楼的柱子由一层石膏板包裹，不仅美观大方，也使空间更加通风、流畅。图书馆的材料和文件都是开放的，这样可以鼓励员工随时阅览，为自己充电。

总而言之，办公室一方面引用了许多FKA公司老式结实地面的趣味和悠闲感，同时还加入了新的复杂感和酷感。就像所有FKA的作品一样，公司的办公室就是一个独一无二的环境，它能告诉人们谁在这里办公，而不是这些人刚好在此处。

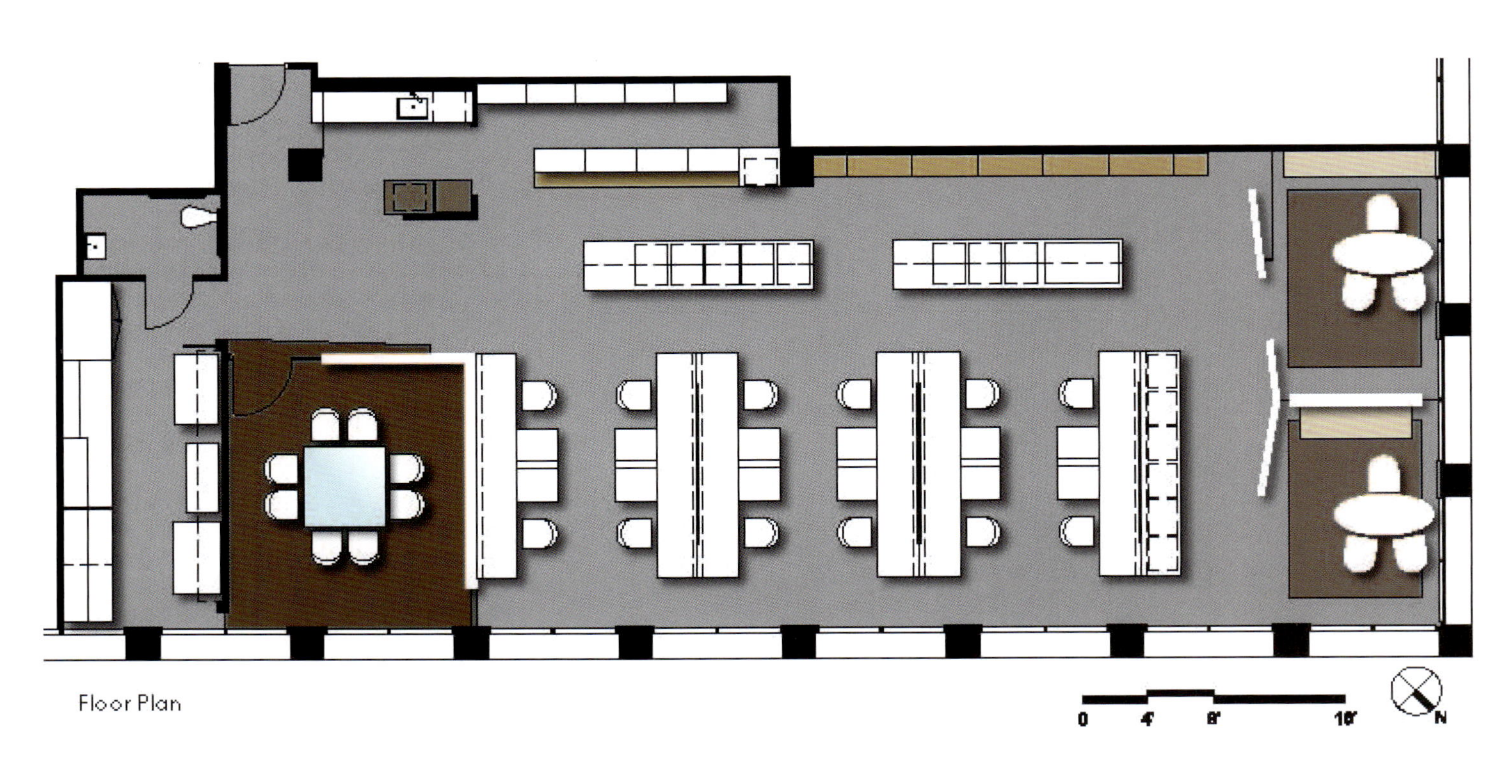

Floor Plan

Designer Gonzalo Mardones Viviani

GONZALO MARDONES VIVIANI ARQUITECTOS

Lighting
Paulina Sir

Constructor
Horacio Rodríguez

Furniture
Orlando Gatica, Mario Galdames

Area
500m^2

Photographer
Guy Wenborne

As architecture studio Gonzalo Mardones Viviani Arquitectos were attracted with the idea to work in the Vanguard Building, their last Offices project. This building, as all their works was created with only one material, in this opportunity with white steel. This material was used in all exteriors and interiors of the building, also in the six storey hall: the soul of the building where the white steel was neutral to permit the light and the total space were protagonist.

As the same way the studio was entirely white. Place in the first sub floor was an L shaped turn a patio of 11m x 11m that is the extension of their atelier and offices. Located in the first sub floor (in a 500 sqm place) permit them to project the studio in a space with an interior height of 330cm with very success natural light. The plan was not orthogonal, that permit as to tense the circulations and the space.

The main idea of the studio project was work in the better spatial and natural light conditions.

As the same way that all their projects and the same way of the Vanguard Offices Building they work with one material and one color, they opted for the white: all the walls and roofs where white. The floor was a light grey ceramic that collaborate with the white sensation (the same floor in the entire studio) and they separate the spaces with transparent crystals that collaborate too. The white permit light and space is the protagonist. The furniture was white too and also the chairs were white, there are no elements that compete with the white, finally with the white they permit the light the most important material.

美诺建筑事务所的上一个设计项目——美诺建筑工作室，是大家想往的办公圣地。像他们的其他设计作品一样，该建筑设计也只采用一种材料——白钢。建筑的里层外表以及六层楼高的大厅用的都是这种材料。被喻为“大楼之魂”的大厅，被白钢的素雅色泽反射的光亮照耀得绚烂夺目。

工作室也顺应总体色调，整体采用全白色。地下层建在一个L形的斜坡上，设计师将那里设计成了一个面积为121平方米（11米 x 11米）的露台作为画室和办公室的外延。在一个面积达500平方米的地下层上建地基，工作室的室内高度可以达到3.3米，并能保证充足的阳光照射。原先的方案因没有采取垂直设计，使空间紧张并限制了其流动性，缺乏可行性。

工作室设计的主旨是让员工享受到更宽敞的空间，沐浴更自然的阳光。

与美诺建筑事务所的其他所有项目设计和美诺办公楼一样，都是采用一种建材、搭配一种颜色，这次他们选择了白色：白色的墙壁和白色的天花板。浅灰色瓷砖铺成的地面与白色（工作室地板的颜色）的视觉效果相呼应。用来分割空间的晶体立面也与灰色瓷砖相映成趣。白色的反射效果增强了光线亮度，也突出了整个空间举足轻重的地位。家具无一例外也都是白色，包括椅子。白色是主导色，有了它，光成为了此次设计中最重要的材料。

Designer Alessandro M. Pierandrei, Fabrizio M. Pierandrei, Stefano Anfossi

THINK GARDEN

Design Company
Pierandrei Associati

Collaborator
Luis Felipe Bueno, Sara Carlini, Elio Di Carlo, Radim Tkadlec

Location
Via Savona, Italy

Area
600m^2

We are all familiar with the definition of "non-place", but what we do not understand is how this concept has been applied so well to the idea of "business lounge".
"Think garden", the first "creative business lounge", is conceived with the idea of not renouncing to propose an experiential space as a place for business meetings and more.
The 600 square meters of space "Ermenegildo Zegna" in Via Savona are then interpreted by the practice Pierandrei Associati, such as an inner wood, complete with tall trees and shrubs, flowers and lawns which run on small artificial hills.
A primordial space, yet friendly and welcoming, with its organic shapes that create meeting places and discussion, relaxation areas and exposure, ensures privacy when needed and blends indoors and outdoors. A place to explore in search of new relationships and new adventures where the appearance of "wild" is mixed with a ubiquitous technology, giving everyone the opportunity to isolate themselves from the hustle of the city without giving up its link with the world.
To harmonize the space use of Beta, the latest innovation from Tecno SpA which was established with the same basis was as the center: a system of office furniture designed by Pierandrei Associati for the creative space that can modulate the intensity of the spaces work and to integrate the different needs of sharing and privacy.
Dedicated to "do business differently", the space is complete with outdoor areas, bar and a series of themed areas aimed at the personal well-being and leisure temporary and thought space, the visitor will be surprised by the "barefoot area", small green area where you can walk barefoot, or "bird cage", clearing where the presence of natural sounds and hammocks invite you to pause, or many other relaxing moments scattered through the center.
"Think garden" is therefore an area dedicated to work, but according to a new philosophy, which is to be in harmony with the environment around us, to combine professional work and relax in a unique experiential space where we can say that doing business is natural.

beta

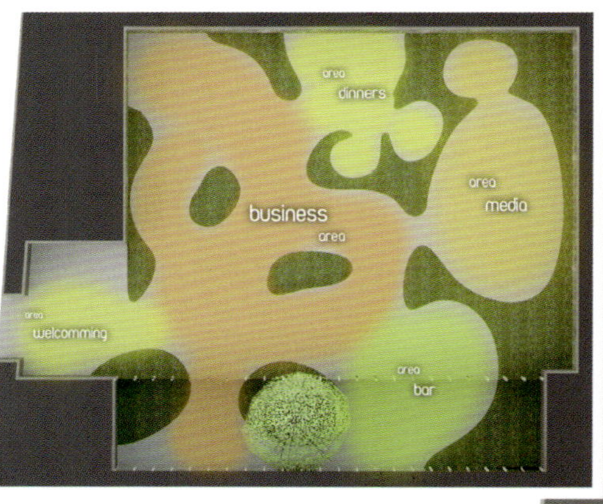
area
dinners
area
media
business
area
area
welcomming
area
bar

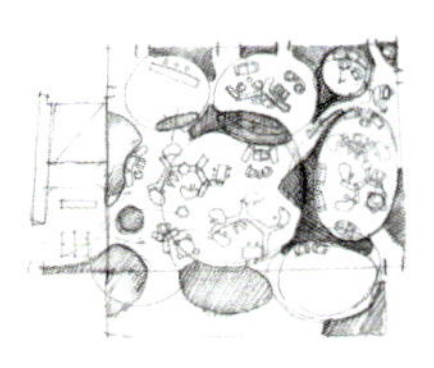

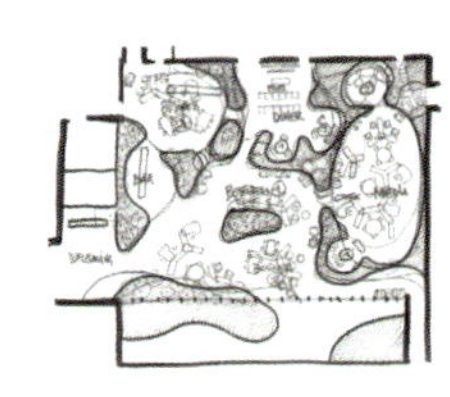

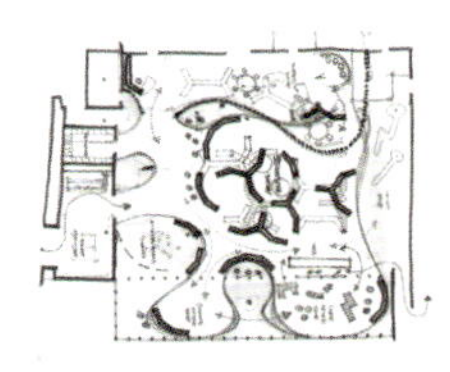

plan 1

① area verde a prato con betulle
② pavimento esistente
③ nuove pareti in cartongesso
④ telo per presentazioni
⑤ rivestimento pareti con grafica
⑥ barriera verde
⑦ tabellone progetto
⑧ panelli diners
⑨ schermi lcd

a soffitto cerchi in barrisol
retroilluminato

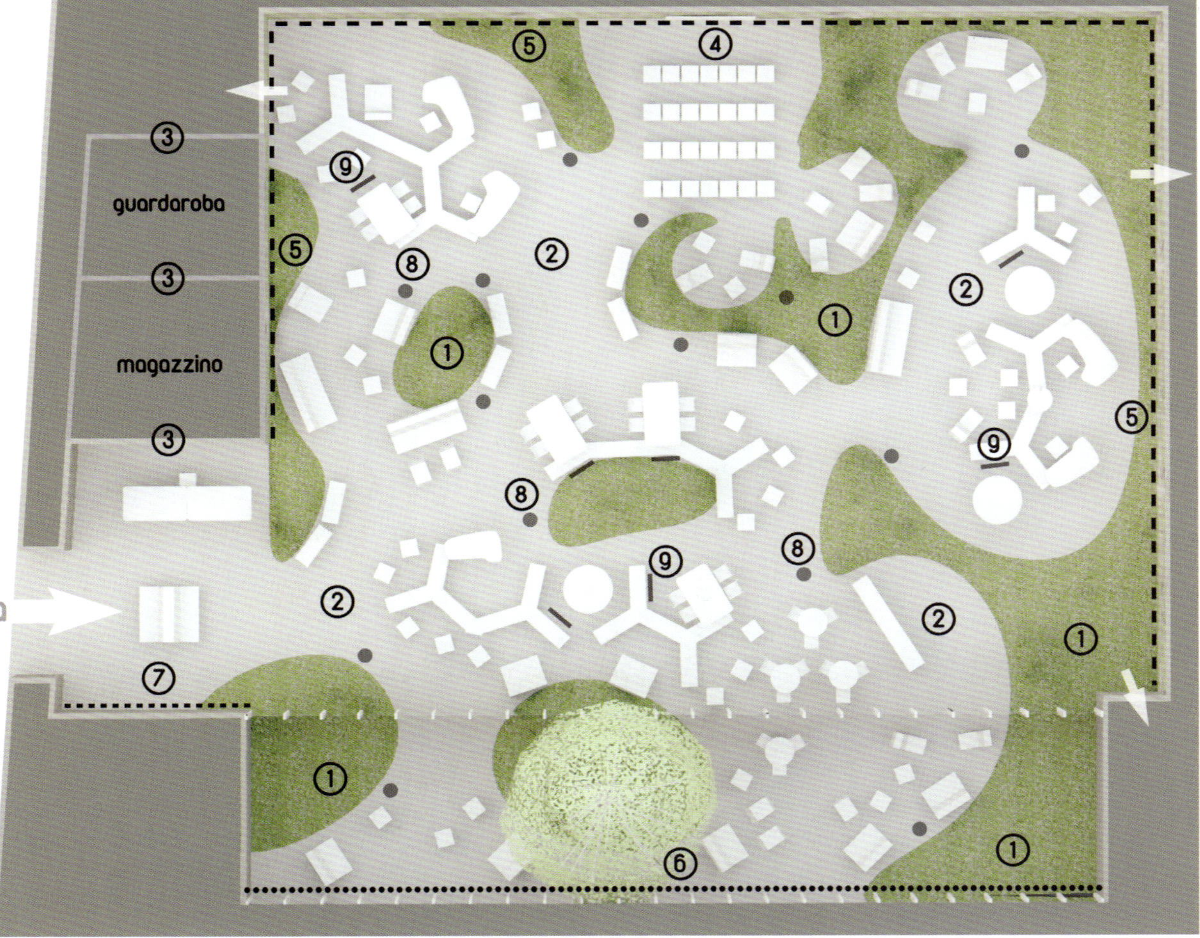

plan 2

我们都很熟悉“非地方”的定义，只是我们不知道这个概念是如何恰如其分地体现在“公司休息室”的设计中的。

“思索公园”被称作第一个“有创意的公司休息室”，它源于坚持将体验式空间打造成商务会谈之地的想法。

Pierandrei Associati 重新诠释了位于萨沃纳的 600 平方米的“杰尼亚”场地，比如建造了内部树林，栽种了高大的树木和灌木，在小假山上铺上鲜花和草坪。

这是一个原始且充满友好与希望的地方，有机的外形提供了会议、讨论、休闲和社交的场所，在需要时保护你的隐私，并将户内户外融合在一起。这是一个结识新朋友、开始新冒险的地方，它将“野外”与无所不在的科技结合起来，让每个人都有机会在与外界保持联系的同时让自己远离尘世的喧嚣。

为了让公开测试投入使用的效果变得和谐，出自用同样基准建立的 Tecno SpA 的最新创意发挥了作用：在这个充满创意的空间里，由 Pierandrei Associati 设计的一系列办公设施能够调节办公室内的工作强度，并将人们对于共享资源与个人私隐的需求整合起来。

设计师致力于营造“做事与众不同”的办公空间，该区域配备了户外区、酒吧和一系列针对个人幸福感、短暂的空闲和参悟等有特定主题的场所，来访者会感到惊喜，只为一处能够赤脚散步的绿地——“赤足区”，或是能够领略到自然风光、在吊床上小憩片刻的“鸟笼”，又或是其他许多散布在中心区域的休闲项目。

“思索公园”虽然是办公专用的场所，但若参照一种新的观念——人与身边的环境和谐共处，将专业办公与休闲融为一个构造独特的空间。那么，在这里我们就可以说：“工作是自然的。”

Designer architect Zane Tetere, designer Elina Tetere

INSPIRED OFFICE

Design Company
SIA OPEN ARCHITECTURE AND DESIGN

Executor
metal artist Edgars Spridzans,
carpenters "LB construction"

Location
Terbatas street 30, Riga, Latvia

Area
380m^2

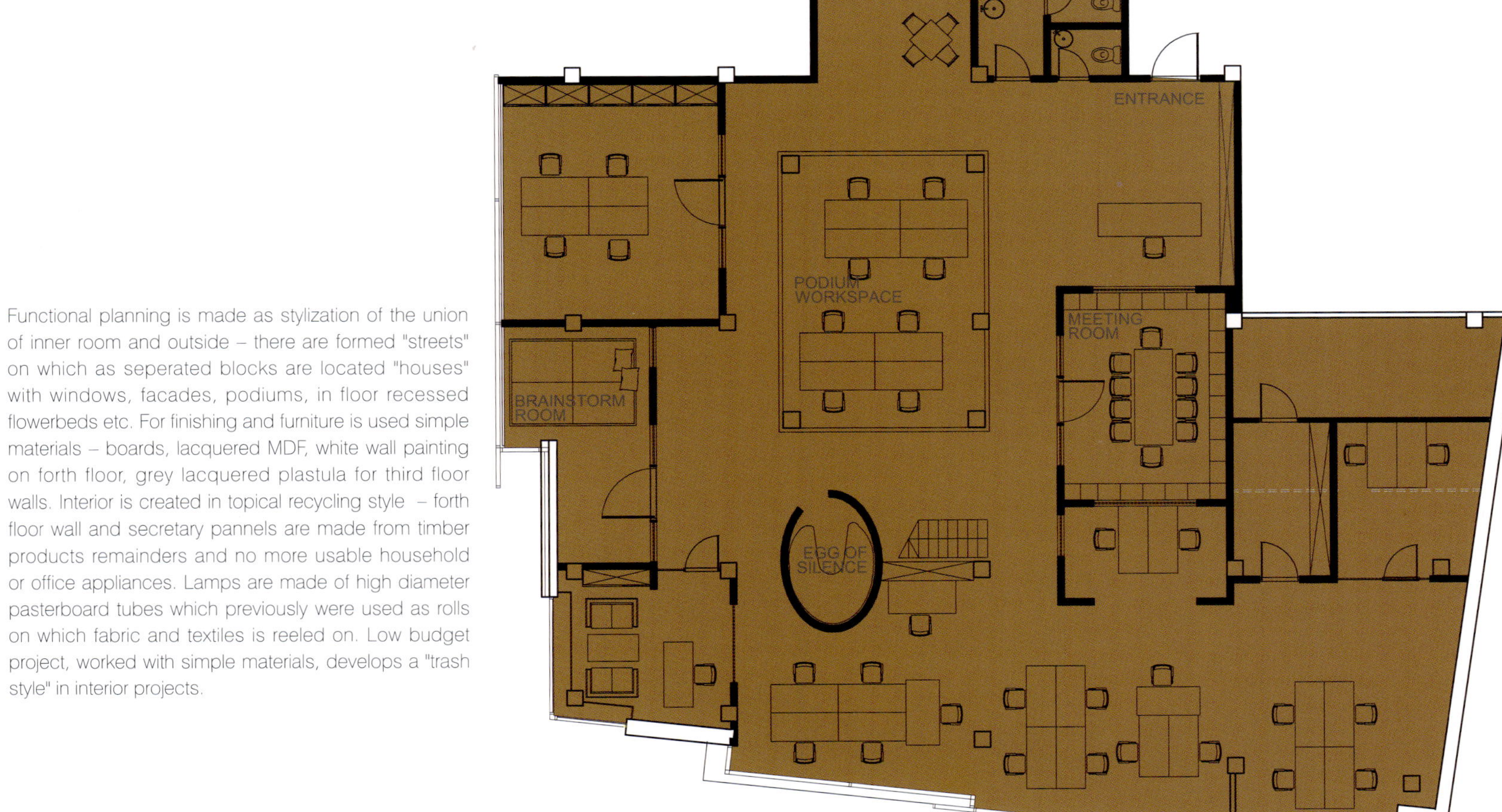

Functional planning is made as stylization of the union of inner room and outside – there are formed "streets" on which as seperated blocks are located "houses" with windows, facades, podiums, in floor recessed flowerbeds etc. For finishing and furniture is used simple materials – boards, lacquered MDF, white wall painting on forth floor, grey lacquered plastula for third floor walls. Interior is created in topical recycling style – forth floor wall and secretary pannels are made from timber products remainders and no more usable household or office appliances. Lamps are made of high diameter pasterboard tubes which previously were used as rolls on which fabric and textiles is reeled on. Low budget project, worked with simple materials, develops a "trash style" in interior projects.

POLICIJA

apollo
european hit radio
Zelta Zivtina
Diena

本案的功能设计因袭了内部房屋与外部空间结合的风格：构造出的“街道”上分布着“房子”，“房子”内有窗户、外立面、矮墙和嵌地式花坛等。而装饰和家具则采用了简单的材料如木板、喷漆纤维板、四层的白色墙面喷漆和三层的灰色喷漆墙面。室内局部采用了现下流行的循环利用风格——四层的墙体和文案板由木制产品的余料和废弃的家用或办公用品制成。灯具则是由以前用来卷织物的大直径石膏管制成。本案以较低的预算，辅以简单的材料，发扬了“变废为宝”的设计风格。

Design Company ONG&ONG Pte Ltd

ONG&ONG SINGAPORE OFFICE

Team Director
Teo Boon Kiat

Architecture
ONG&ONG Pte Ltd

Project Management
O&O Project Innovations

Location
510 Thomson Road, Singapore

Photographer
See Chee Keong

Spanning the ninth floor to the eleventh floor, the ONG&ONG office was in need of a revamp to integrate the three levels.

The heart of the interior design concept lies in the suspended staircase that traverses all three floors, facilitating movement between these. Extra space around this core structure houses shared facilities such as the reception area, recreation and meeting rooms.

Industrial chic was the desired overall look, with bare floors and exposed ceilings giving the office an air of casual sophistication. Echoing ONG&ONG's corporate color, splashes of orange are playfully interspersed between white walls – this maintains the corporate atmosphere whilst lending an air of ebullience for a healthy work – life balance.

ONG&ONG

EXIT

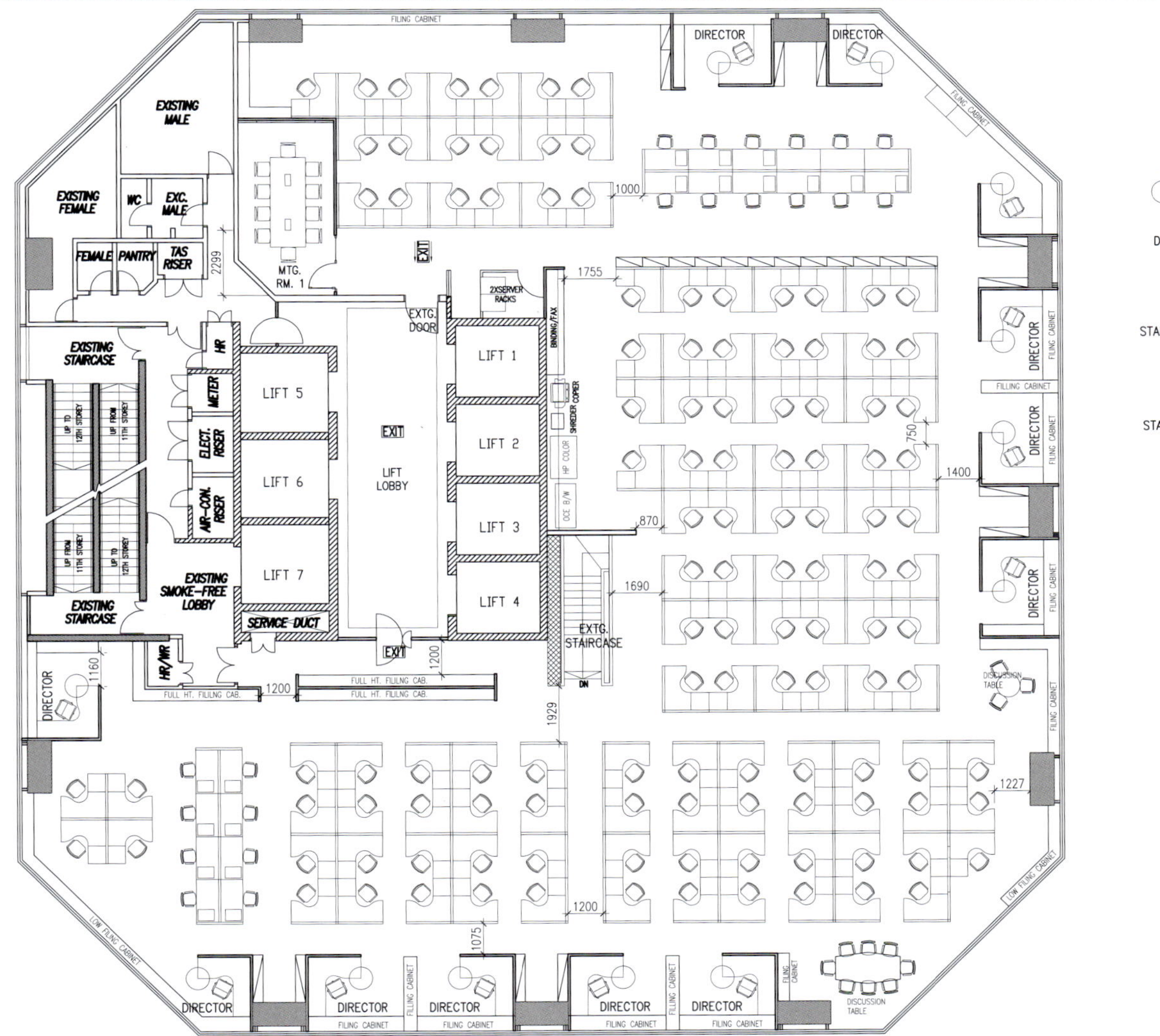

7TH STOREY FURNITURE LAYOUT PLAN SCALE 1:150

DIR = 12 PAX

STAFF = 20 PAX

STAFF = 122 PAX

从九层跨越到十一层，ONG&ONG 公司的确需要改建一下来把这三层整合起来。

该室内设计的核心在于纵贯整个三层楼、使上下移动便利的悬浮楼梯。围绕核心建筑构造的额外空间则是一些共享设施，例如接待区、娱乐室和会议室。

设计的别致之处体现在精心设计的整体外观上，不加装饰的地板和天花板给办公室带来一种休闲的感觉。泼溅式的橘色趣味地点缀在白色墙面之间，呼应了 ONG&ONG 的企业色调。这在保持公司气氛的同时也为工作与生活的健康平衡带来了一丝活力。

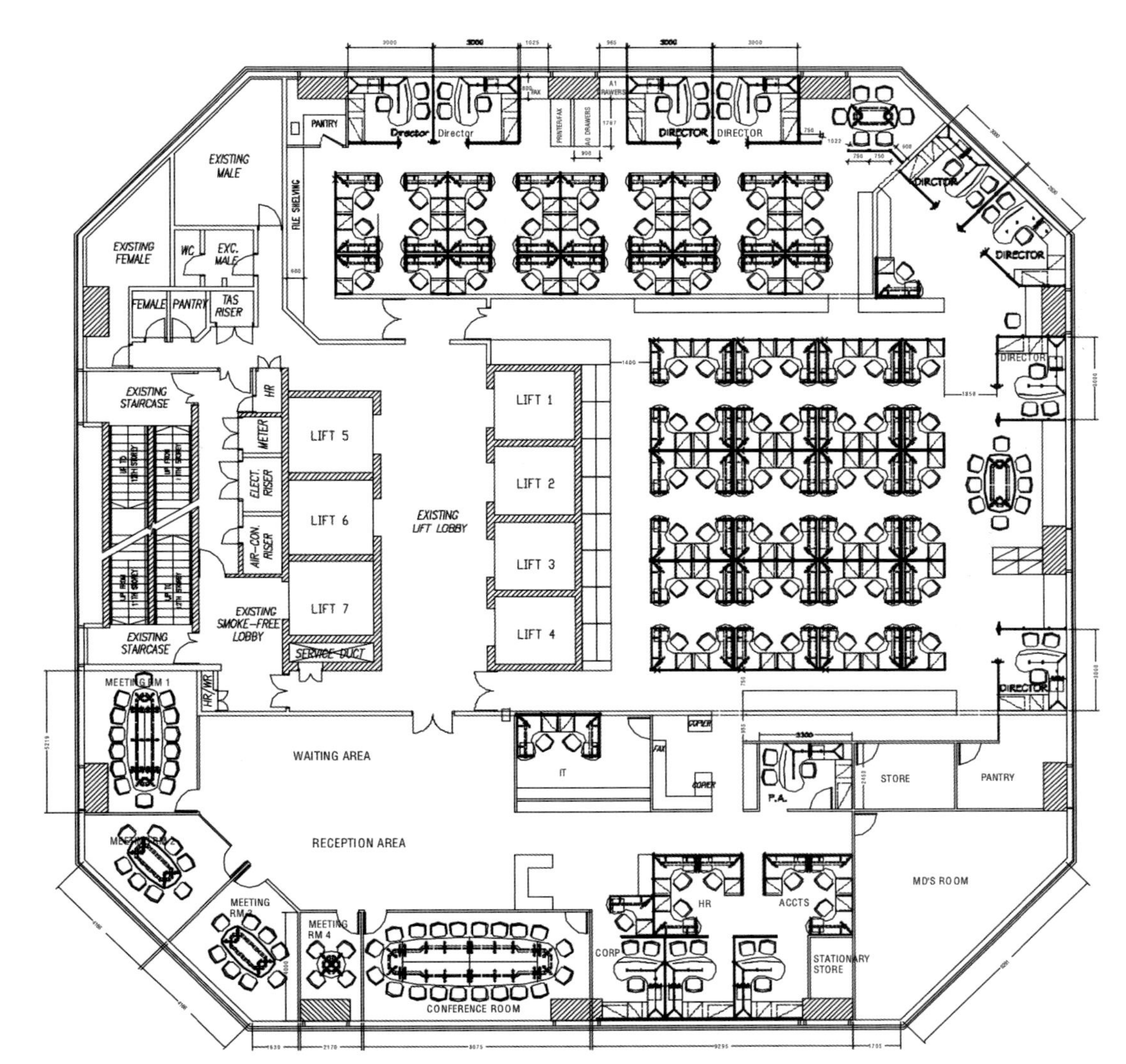

MD	-1 ROOM
P.A.	-1 SEAT
ACCTS	-4 SEATS
HR	-4 SEATS
CORP.AFFAIRS	-2 SEATS
IT	-2 SEATS
MEETING ROOM	-4 ROOMS
DISCUSSION TABLE	-2 SETS
DIRECTORS	-8 CUBICLES
ARCH/TECH/CSC/VISUAL	-69 SEATS

9TH FLOOR

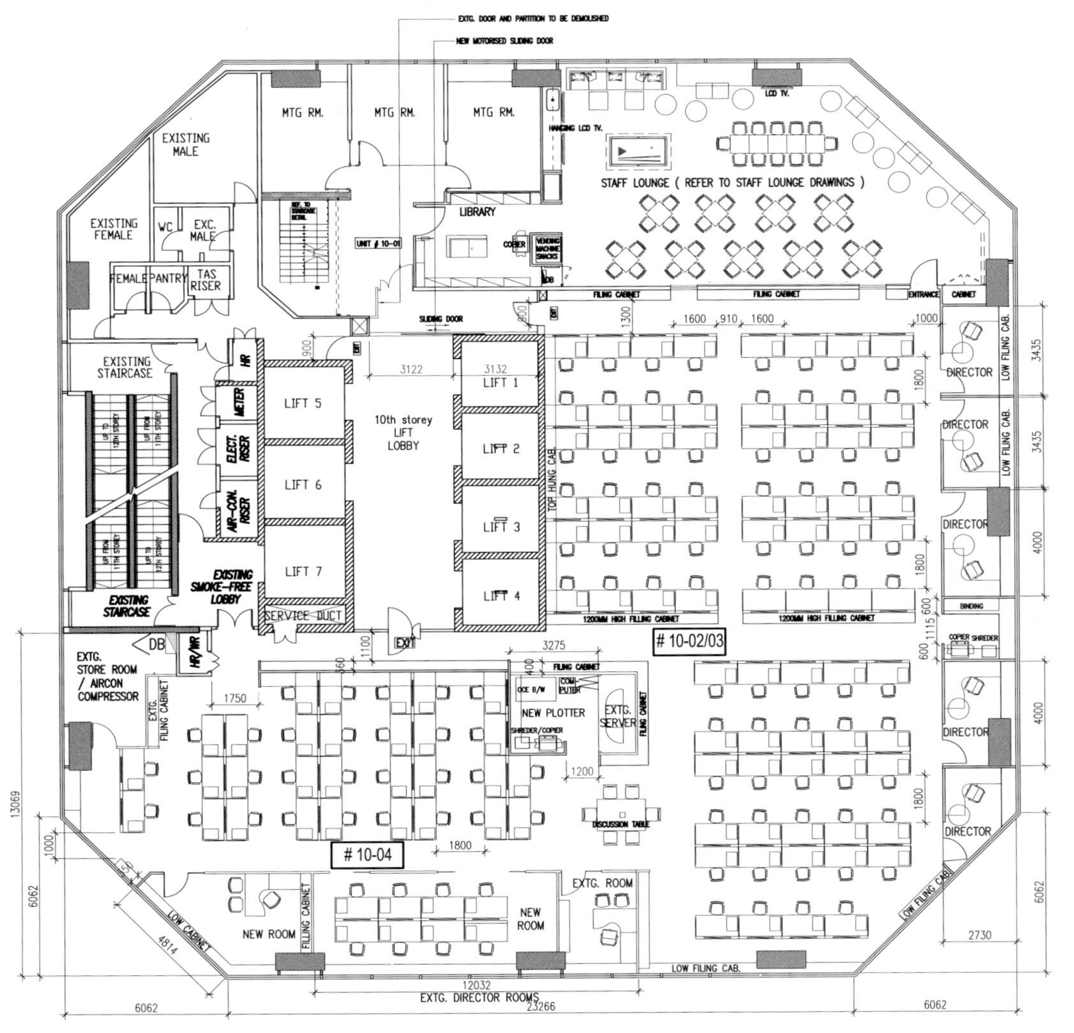

DESCRIPTION

(TOTAL WORK STA. = 107 nos.)

(TOTAL EXTG. DIR. ROOMS = 5 nos.)

(TOTAL NEW DIR. ROOMS = 5 nos.)

EXTG.

10TH STOREY FURNITURE LAYOUT PLAN

SCALE 1:150

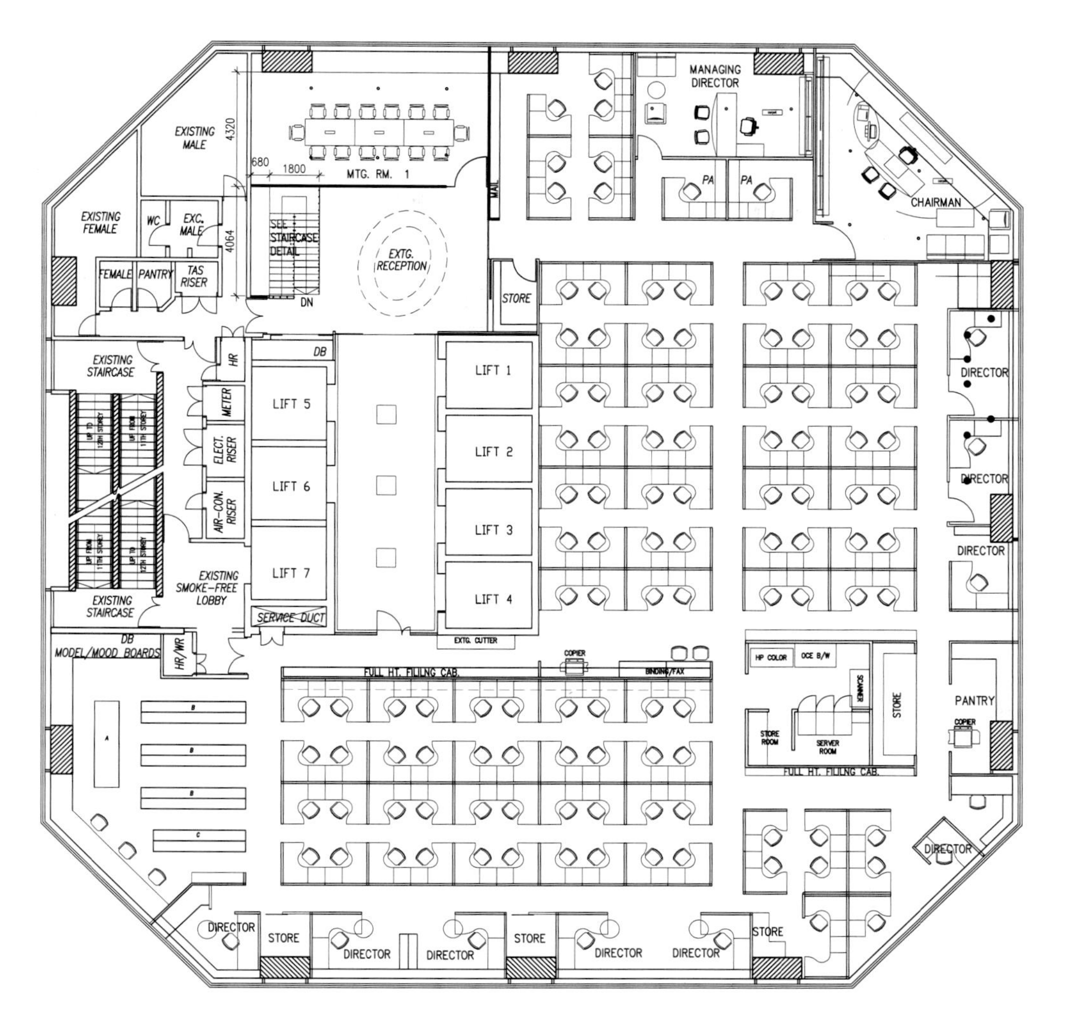

11TH STOREY LAYOUT PLAN

SCALE 1 : 100

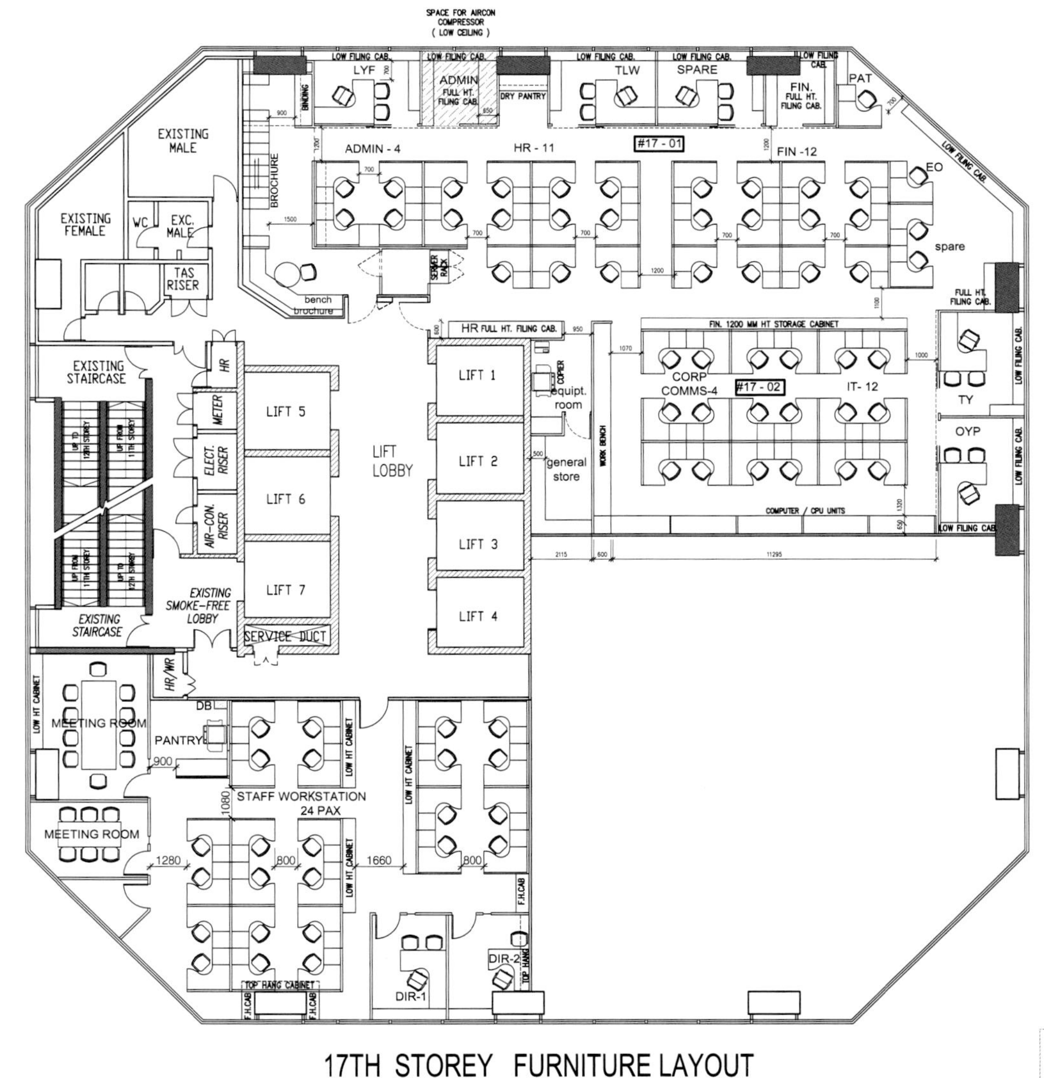

17TH STOREY FURNITURE LAYOUT

AS-BUILT

Design Company Sub Estudio

SANTA CLARA AD AGENCY

Location
Vila Madalena, São Paulo, Brazil

Photographer
Fran Parente

Located in São Paulo, in Isay Weinfeld's new building, the Ad Agency Santa Clara was designed by Sub Estudio from the architecture project to the furniture.
Occupying the middle of a street block, the building consists in two different parts, a square, that is right on the middle of the street block, and a real long rectangle that makes the access to the street.
The Agency also needed to be divided in two different pieces. On one piece, they had to be more flexible: the creation area. They need to change the tables' layout from time to time to serve the client better with his special needs.
On the other piece was the edition area, the traditional work area that doens't have to change.
On the square block, it's located in the creation area. A big open space where the tables were designed to be a single work base that can move around. While in the long rectangle only one table was projected and goes through the rectangle making a more rigid layout.
Both of the blocks have a really high height that made it possible to have a second floor where located all the rooms that had to be closed, such as computer room, administration room, and meeting rooms on different sizes.
Whenever possible, the meeting rooms where designed with glass walls allow a nice natural light through the circulation area. Since the project just use part of the higher floor, all the meeting rooms have a nice view to lower floor or to the neighborhood outside.

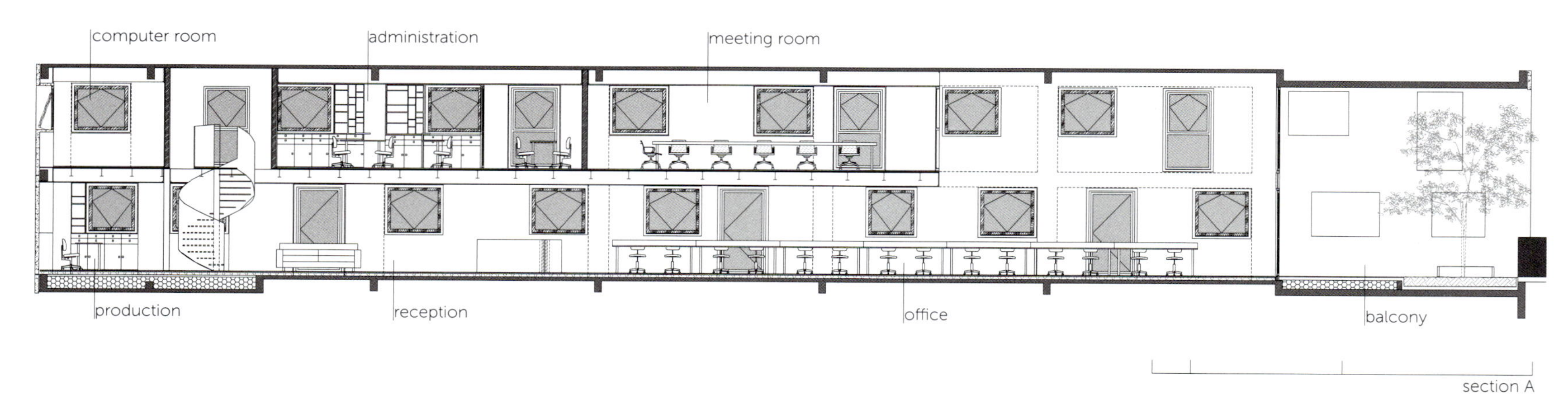
computer room
administration
meeting room
production
reception
office
balcony
section A

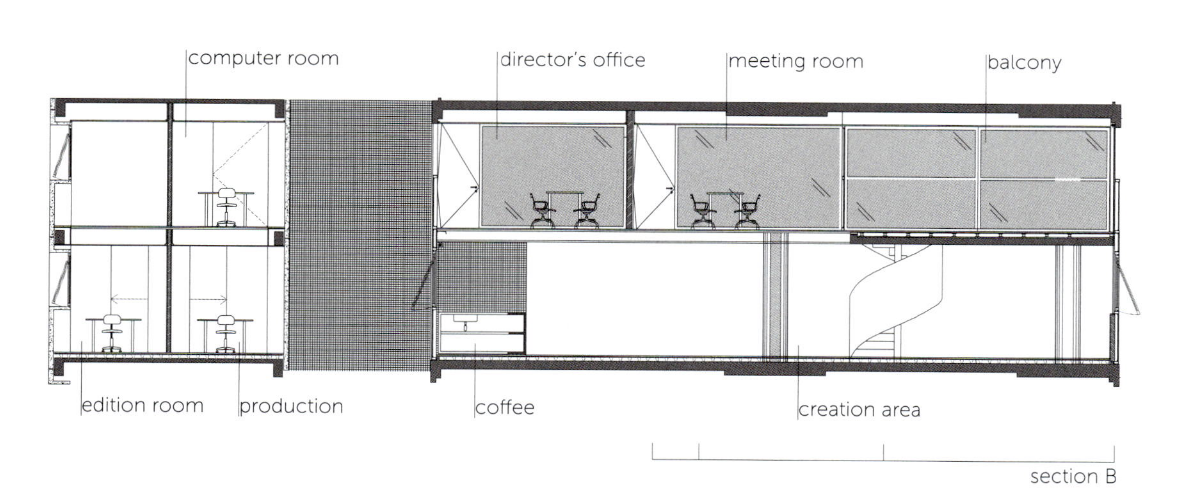
computer room
director's office
meeting room
balcony
edition room
production
coffee
creation area
section B

Pronto
Hora.

圣塔克拉拉广告公司坐落于圣保罗，在伊塞·魏因费尔德设计的一栋新建筑里，该广告公司从整体结构到家具均由E工作室的分部负责设计。

这栋占据了街区中央的建筑由不同的两部分组成：一部分是正方形区域，正好在街区中央；另一部分是一个十分狭长的矩形区域，紧挨着街道。

广告公司也需要分成不同的两块。一块是需要更多灵活性的创意区。他们需要不断地改变桌子的布局以更好地服务于客户的个性需要。

另一块是编辑区，一块无需变化的传统办公区域。

创意区位于正方形区域，这里是一个很大的开放空间，桌子的布局是在可随处移动的基础上为单独的工作设计的。而狭长的矩形区域内只有一张桌子，布局相对固定。

两个区域都有充足的空间高度，可以建设第二层楼来安置封闭的房间。例如，电脑室、行政办公室和大小不一的会议室。

会议室在任何可能之处都设计了玻璃墙，使自然光线随时可以照进流通的区域。由于设计方案只占用了二层的部分区域，所有的会议室都可以直观地看到一层和附近室外的情况。

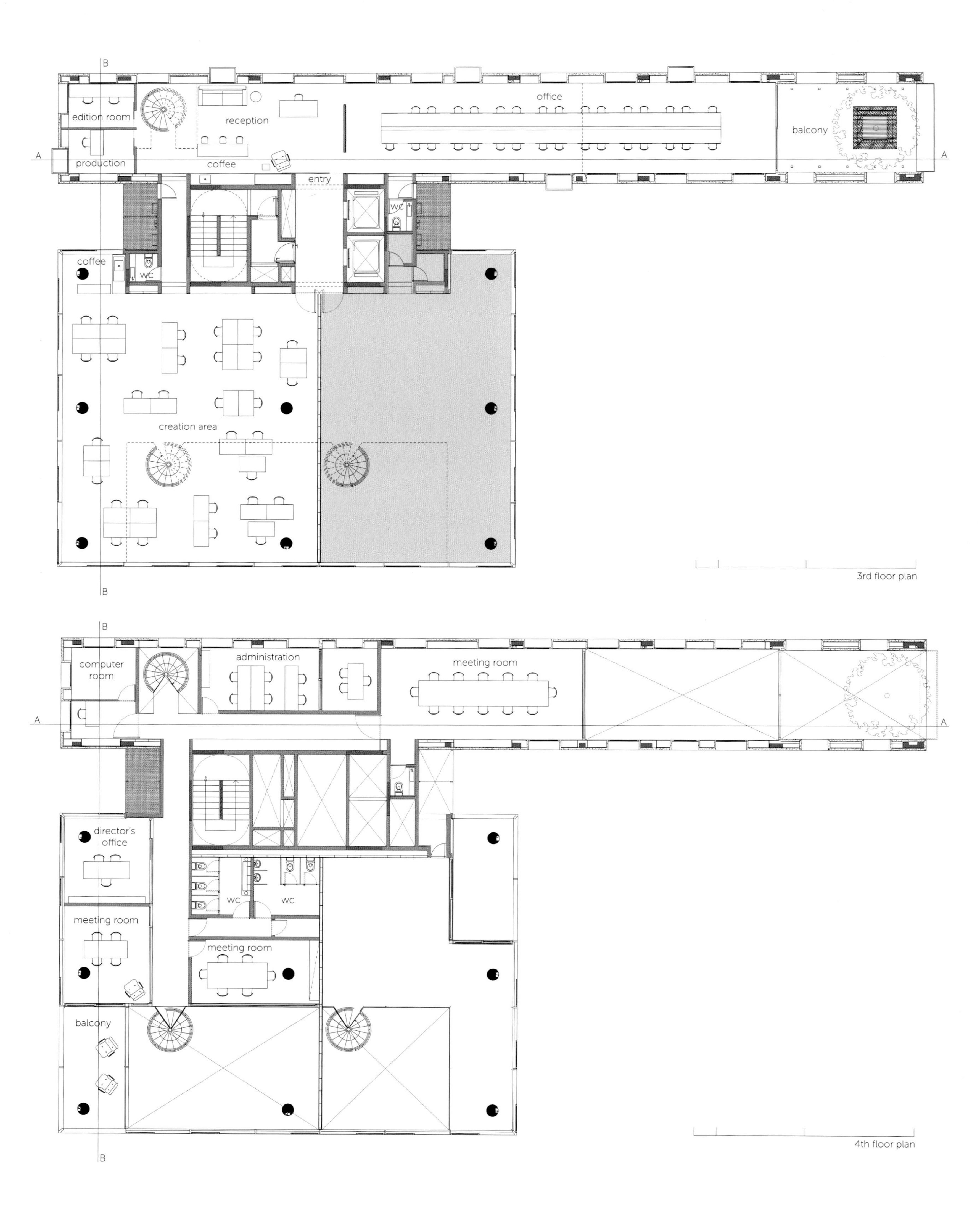
B
edition room
reception
office
balcony
A
production
coffee
entry
wc
coffee
wc
creation area
3rd floor plan
computer room
administration
meeting room
director's office
wc
wc
meeting room
meeting room
balcony
4th floor plan

CLARA

Design Company Liveperson Headquarters

LIVEPERSON DUE DILIGENCE

MEP Engineer
JFK&M consulting engineers

Location
475 10th Ave (at 36th), New York, USA

Area
1,394m^2

Liveperson, an online marketing, web analytics, and expert advice company, asked Mapos to design their new corporate headquarters in a way that was sustainable.

A critical first step was the co-authoring of the program through a custom-designed game. Laid out on their office floor, the game board featured program "pieces" that represented standard office needs. It also featured "wild cards" which could be put towards any use they desired. Mapos refereed a boisterous session where different departments haggled and consolidated, traded and evaluated certain pieces for others. The result was an approved brief – they could all build upon that included a public "square", informal kitchen that would be a part of reception, work space for 80 (yet flexibility for 100), several small meeting rooms, a game room, and a few large meeting rooms. The office is centered on a large "town square" that is adjacent to the open kitchen and visible from the entry and reception area. Next to the town square is a shingle style "building" containing two large meeting rooms. These rooms can be used individually, be combined to form a larger meeting room, and even opened up to the town square to completely remove all the walls in the center of the office. The rest of the space is essentially an open desk layout interspersed with small stand-alone meeting rooms. Each of these rooms is completely sound-proof for private meetings and conversations and bears a "crown" of signage and greenery.

Another key component of the design was the overall strategy of re-use. All of the materials left by the previous tenant were mined and catalogued to become the building materials for the new space. The existing carpet was brought up and the original concrete floor was cleaned and sealed, while the carpet tiles were used as sound insulation in certain new partitions. The existing light fixtures were removed and re-installed in clustered chandeliers in the reception and large meeting rooms. All of the existing wood shelves, desks, and millwork were disassembled and used for new desk construction. The shingles on the exterior walls of the central meeting room are a combination of wood materials cut down to the same size and applied as lap-siding. The sculptural walls in the reception area are made out of this re-used wood as well and composed to display the accolades, awards, and mementos that tell the history of Liveperson.

XML

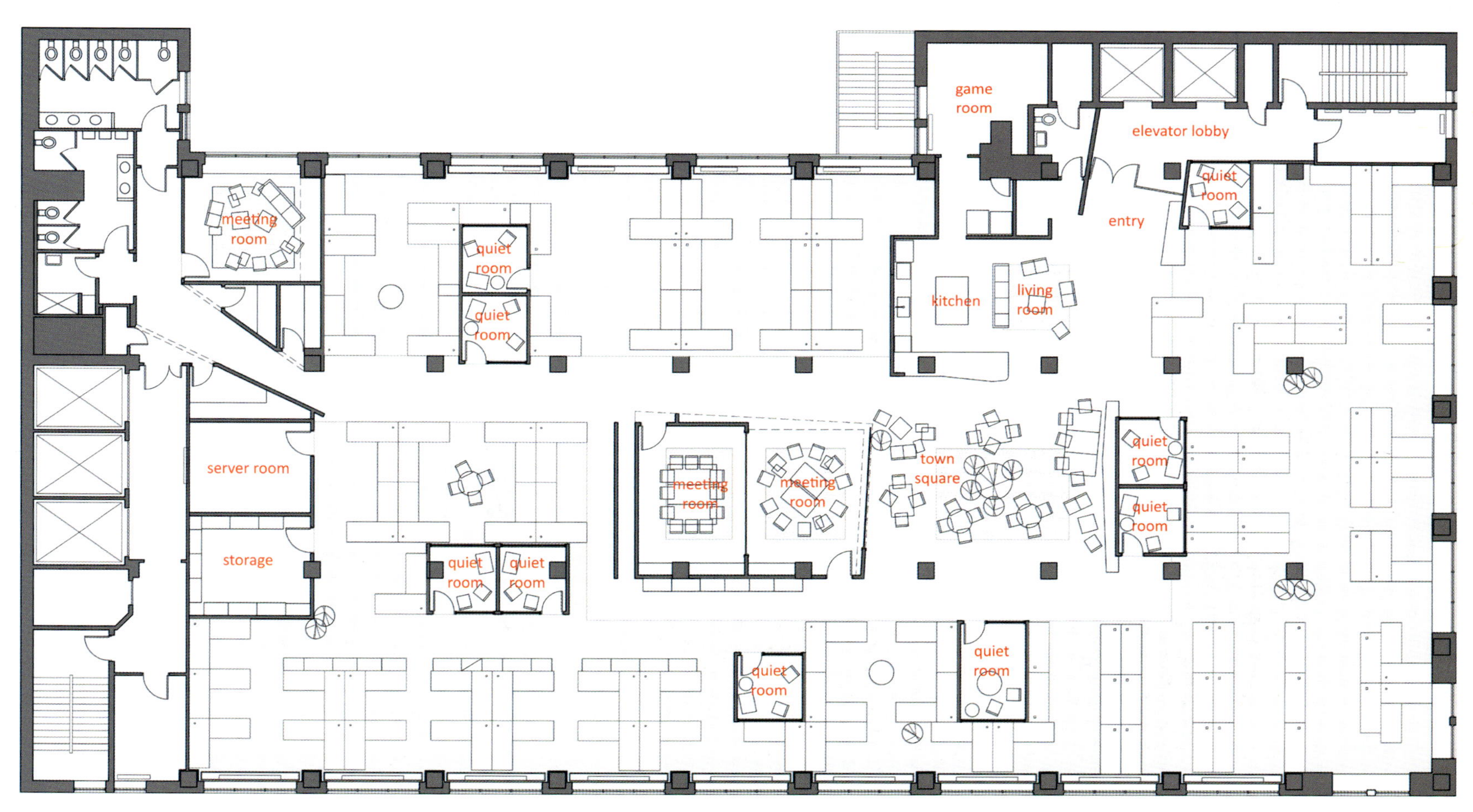
game room
elevator lobby
quiet room
entry
meeting room
quiet room
quiet room
kitchen
living room
server room
town square
quiet room
quiet room
meeting room
meeting room
storage
quiet room
quiet room
quiet room
quiet room

Plan notes

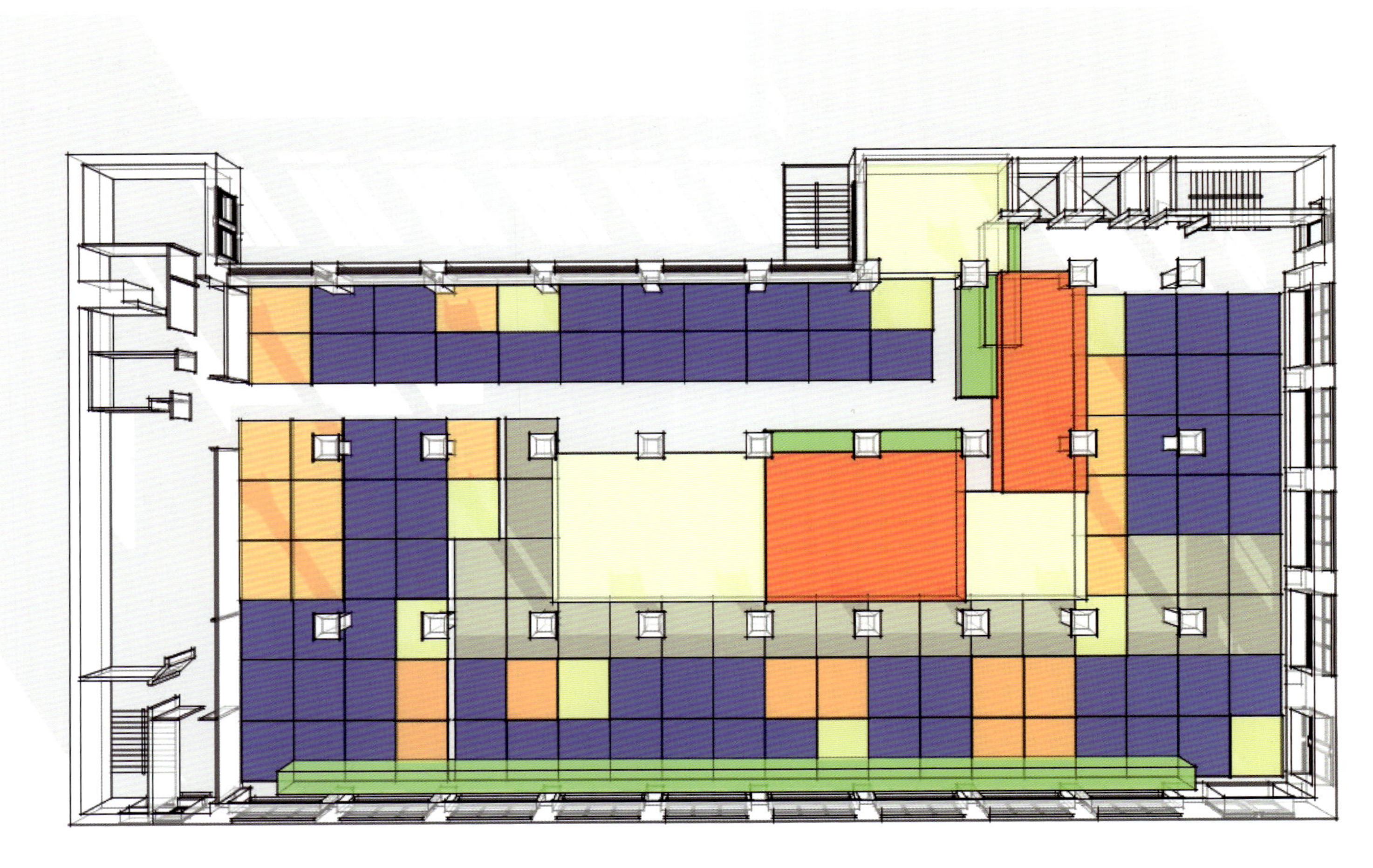

Plan game

Liveperson 是一家以在线营销、网络分析为主营业务，为顾客提供专业意见的公司。该公司委托 Mapos 以可持续发展为宗旨设计它的新总部。

关键的第一步是通过一场客户设计比赛，让客户参与到设计方案的讨论中。比赛在客户的办公场地举行，游戏板上的“牌”代表标准办公室需求；“万能牌”则有助于他们发挥自己的创造力。Mapos 为这场疯狂的赛事做裁判，公司不同部门间相互争论、妥协，不停地对他人的想法做出探究和评估。经商讨最终得出了一套可行性方案——在办公大楼内搭建一个公共大厅；在接待区里设计一个居家风格的厨房；改造办公区使其可容纳 80 人乃至 100 人；修建若干小型会议间、一间游戏室和几间大型会议室。办公室集中在“公共大厅”，从大楼入口和接待区可以看见它紧挨着开放式厨房。公共大厅旁是一个木瓦结构的“建筑”，里面有两间大型会议室，既可以单独使用，也可以合并成一间更大的会议厅，甚至可以拆下这些会议室内的所有非承重墙板，将整个会议厅完全呈现在公共大厅中。办公大楼的其他空间被设计成了开放式的办公桌布局，办公桌间散落着若干个小型的独立会议间。这些会议间都做了良好的隔音处理，以绿色植物做装饰，玻璃上有王冠的标志，方便员工召开私人会议、进行私下交谈。

该设计的另一个关键部分就是材料再利用的整体策略。前任租户留下来的所有材料，都被规整分类作为建筑材料以打造新的空间。设计师将现成的地毯卷起来，打扫干净地面、填平地缝，在划分出来的新区域里用拼接地毯做隔音材料。另外，现有的照明设备被转移并重新安装在接待室和大型会议室的吊灯之间。原来所有的木质储物架、办公桌和木制品都被拆掉重新组装，用来制作新桌子。中心会议室的外墙面板由重叠的木板打造，每块木板的大小一致，作为互搭板壁。接待区的雕刻墙也是这些木材循环利用做成的。墙壁上的雕刻诉说着 Liveperson 一路走来的荣耀、赞誉和回忆。

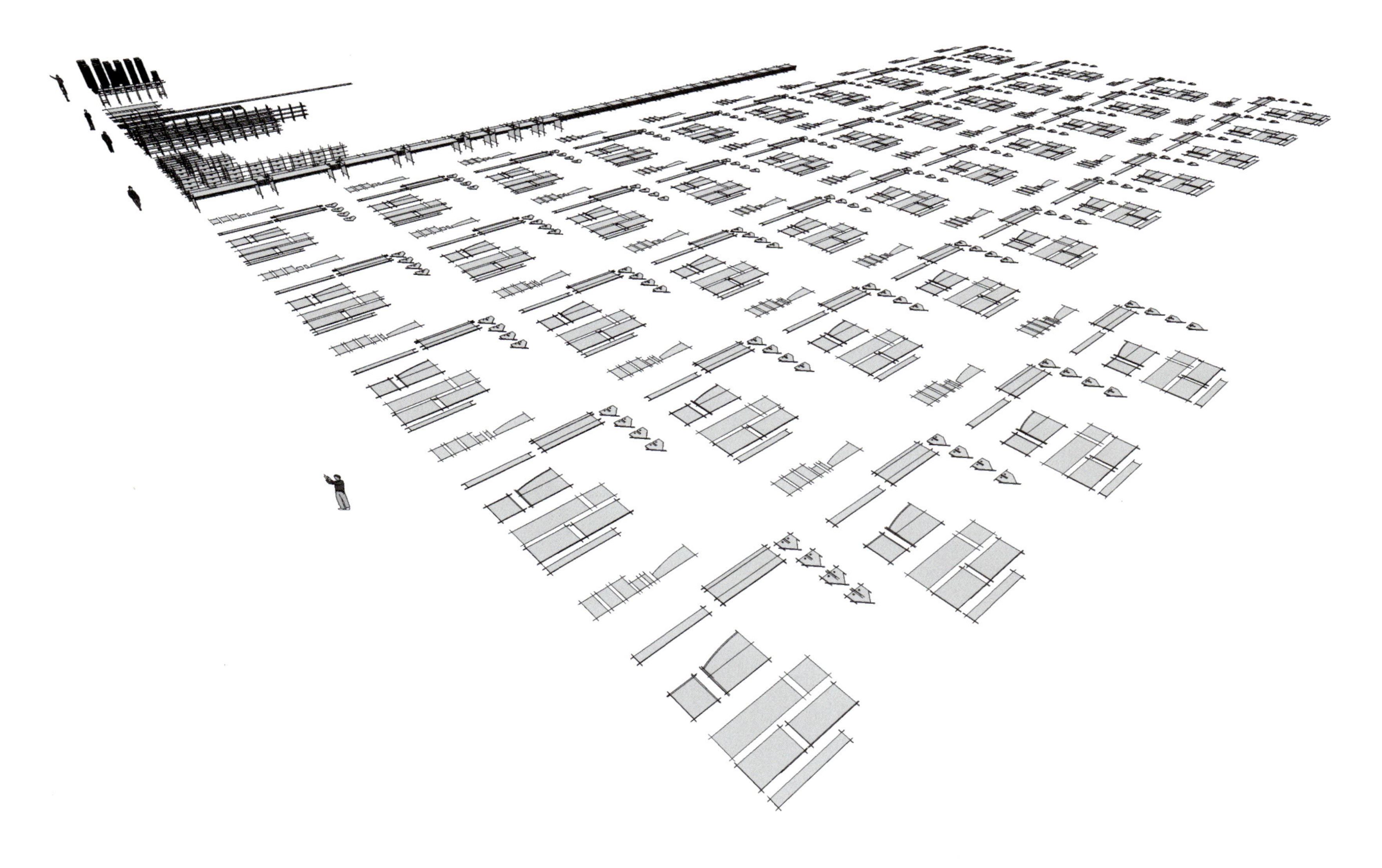

Designer Rosan Bosch & Rune Fjord

LEGO PMD

Design Company
Rosan Bosch Ltd.

Location
Lego System A/S, Billund, DK

Area
2,000m^2

Material
Polyurethane floor, carpet, dry wall, acoustic bats w. graphic print, glass partition walls, acoustic ceiling, furniture

Photographer
Anders Sune Berg

Now, the designers of LEGO's development department, LEGO PMD, has a physical working environment that corresponds to its playful content – a working environment where fun, play and creativity are paramount and where the physical design gives the adults a chance to be part of children's play. With the values "fun", "unity", "creativity & innovation", "imagination" and "sustainability" as basis for the design, LEGO PMD has become a unique development department where the designers can become part of the children's fantasy world.

In order to create a design that corresponds to the focus on play, innovation and creativity, imagination has been given free rein. Across the room, an existing walkway has been transformed into an oversized sitting environment, where a light-blue padding turns the walkway into a light and soft cloud. The cloud unfolds and expands into sofas, sitting space and a slide that connects the two floors in a fun and playful way. The idea of scale is challenged with design elements such as huge grass wall graphics and a giant LEGO man and tables with built-in bonsai gardens, thus playing with perception and scale – who is big and who is small? Where does work stop and imagination start? Through the physical design, the children's fantasy worlds become part of the everyday life, creating the setting for the creation of new design for new games and play. Furthermore, the design of LEGO PMD makes it possible for the designers to work closer together. At ground floor, the open space at the centre of the room creates a dynamic flow where informal meeting places create a setting for social interaction and exchange of information. Towards the sides, there is room for concentrated work, and specially designed means of exhibition such as the show-off podiums and the model towers give the designers a chance to display their work to each other, facilitating the sharing of knowledge and ideas across the department. On the first floor, an expansion of the balcony has made room for five small and three large meeting rooms in each their own color with glass facades and a view of the large, open space. A Fun Zone with a yellow table bar creates room for relaxation and social interaction, where a number of building tables for children make it possible for LEGO's youngest employees to test the newest models and products.

4
3

乐高研发部PMD办公室里的设计师们，如今有了与其有趣的工作内容相呼应的工作环境。这里秉承娱乐、活泼和创造力至上的理念，有形的设计让大人们有机会参与到儿童游戏当中。本案的基本设计理念是"趣味"、"一致"、"创造与创新"、"想象力"和"可持续性"，乐高研发部也因此成为了一个独一无二的研发部门，在这里，设计师们成为孩子们天马行空世界的一部分。

在设计过程中，为了将设计理念紧扣娱乐、创新和创造为主的主题，想象力发挥了不容小觑的作用。设计师们将原本贯穿办公区的走廊改造成一个相当宽敞的休息区，淡蓝色使整个空间看起来如同一朵轻盈柔软的云。云朵舒展开，变成了沙发、休息区和滑梯，滑梯巧妙地连接了两个楼层。关于大小的概念受到一些设计元素的挑战，如巨大的图表墙纸、乐高巨人和自带盆栽的写字台。我们的认知和元素本身的尺寸，究竟孰大孰小？工作在哪里结束？想象力又从哪里开始？空间设计让孩子们的梦幻世界成为日常生活的一部分，为新游戏的设计创造了有利的环境。不仅如此，乐高研发部的设计也拉近了设计师们的距离。一层正中间的开放空间创造出一种充满活力的动态感，为人们非正式会面提供了社交和信息交换的平台。开放空间四周的区域，可供工作人员进行研发；同时，还提供了一些像讲台和模型塔的特制展示工具，给设计师们一个相互展示各自工作成果的平台，促进了部门内部知识与创意的共享。在二层，由于阳台的扩建，室内整体形成了五小三大的会议室格局，每一间会议室颜色各异，有玻璃墙，视野开阔。娱乐区的黄色吧台营造出一种放松和社交的氛围，这里有儿童专用的写字台，为乐高最年轻的员工们提供了测试最新模型和产品的场所。

LEGO

LEGO

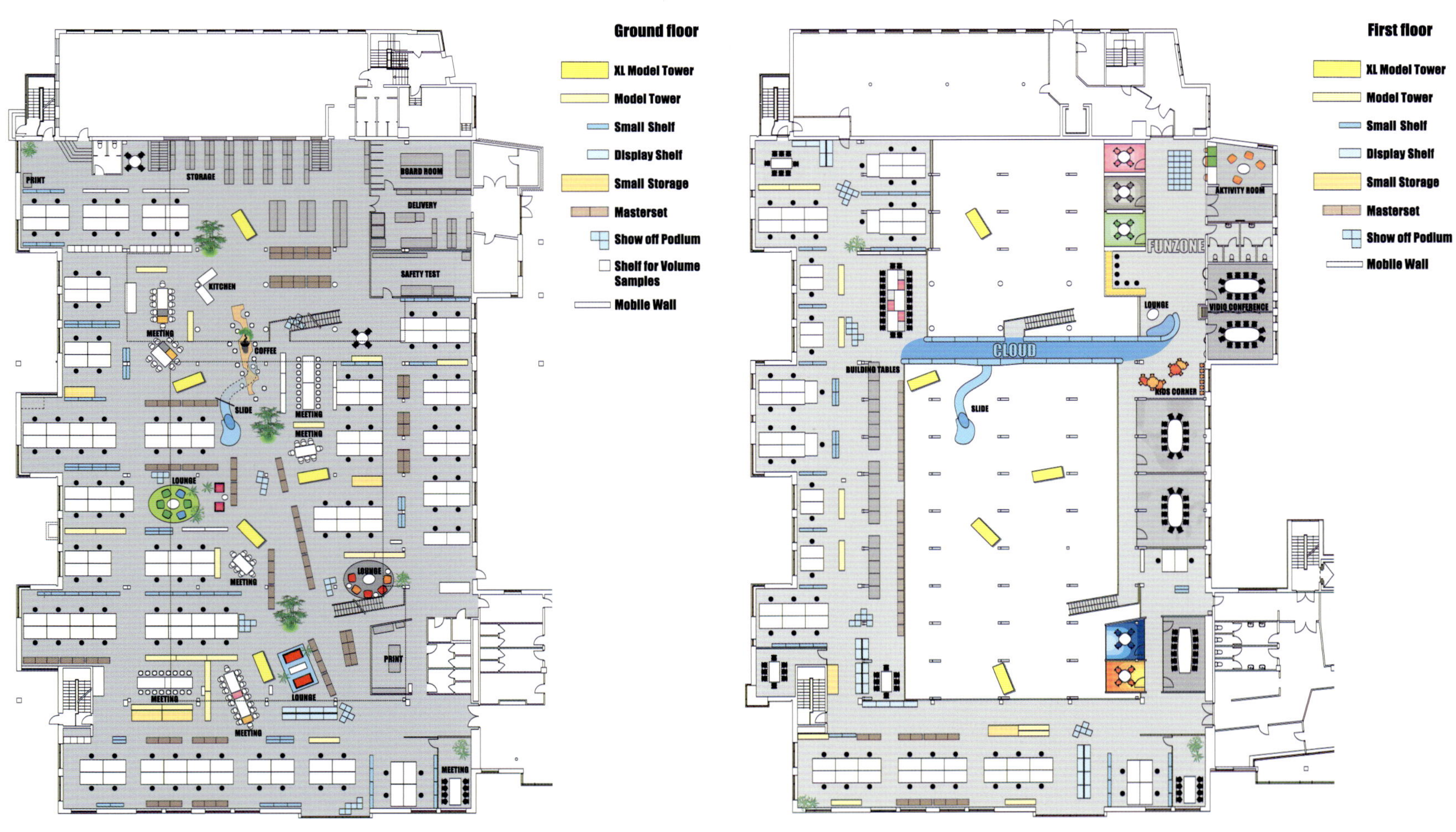
Ground floor
XL Model Tower
Model Tower
Small Shelf
Display Shelf
Small Storage
Masterset
Show off Podium
Shelf for Volume Samples
Mobile Wall
PRINT
STORAGE
BOARD ROOM
DELIVERY
KITCHEN
SAFETY TEST
MEETING
COFFEE
SLIDE
LOUNGE
First floor
XL Model Tower
Model Tower
Small Shelf
Display Shelf
Small Storage
Masterset
Show off Podium
Mobile Wall
ACTIVITY ROOM
FUNZONE
LOUNGE
VIDIO CONFERENCE
CLOUD
BUILDING TABLES
SLIDE
KIDS CORNER

GREY ROOM

STAR
WARS

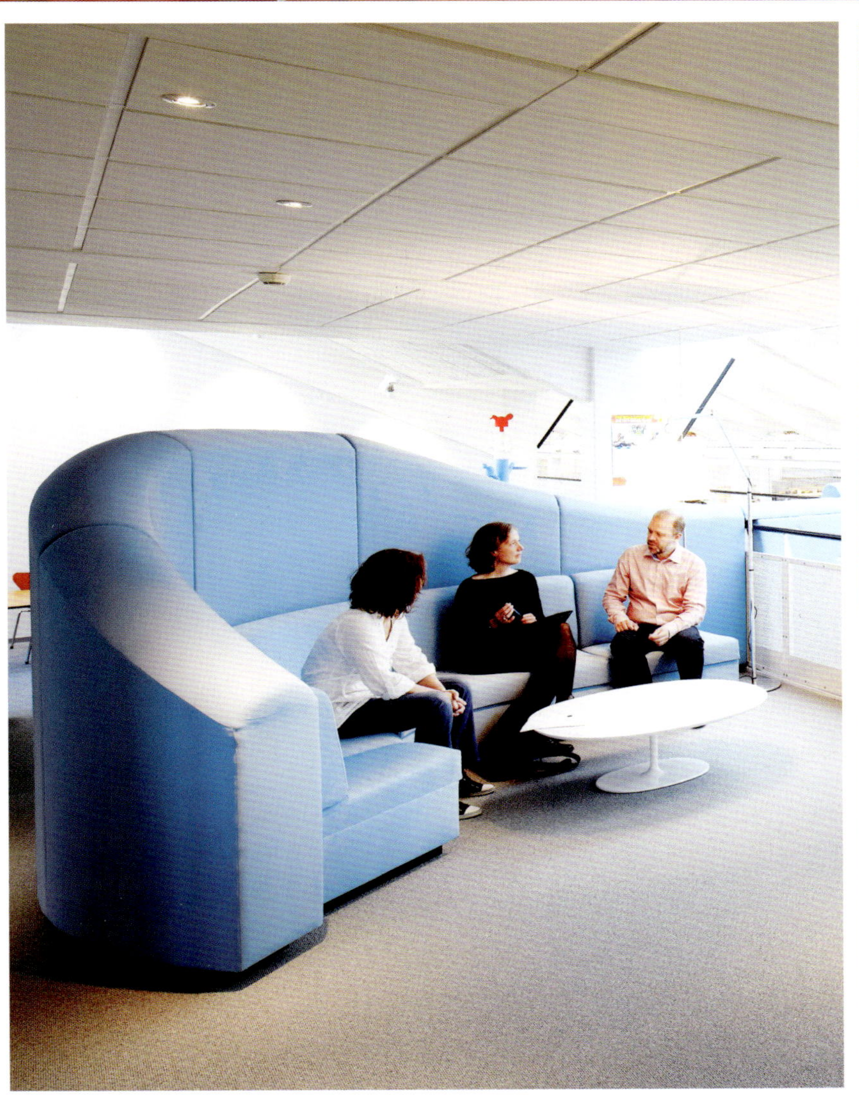

Designer Cary Bernstein Architect

ONE & CO

Area
604m²

Photographer
Cesar Rubio

One & Co is part of a new generation of industrial design firms on the rise in San Francisco. Since its founding 10 years ago, the firm has made its home in a turn-of-the-century brick warehouse in the Mission District. When the need came to expand into an adjacent space, the three principals saw the opportunity to create a workspace that represented the culture of their collaborative studio, design values and the firm's increasing prominence.

At 604-square-meter, the new studio nearly doubles the size of One & Co's original space yet the design maintains enough intimacy to support fluid working relationships among the staff. The open plan takes advantage of the generous windows to the south and east, setting more discreet uses – fabrication room, lounge, "war rooms", kitchen and bathrooms – along interior walls. War rooms can be temporarily occupied by design teams and concealed by curtains when needed to preserve confidentiality.

The minimalist vocabulary of the new construction enhances the character of the older building and provides a range of visual and tactile experiences through essays on translucency, transparency, materiality, mass, texture, color and light. The neutral palette is punctuated by hot pink surfaces on the walls, floor and furniture which enliven the perception of depth. Walnut, used in both the architectural paneling and furniture, adds warmth, naturally bridging both perfect and imperfect worlds of new and old.

SITE PLAN
1 ENTRY
2 RECEPTION
3 SMALL CONFERENCE
4 MATERIALS LIBRARY
5 LARGE CONFERENCE
6 PARTNERS
7 OPEN STUDIO
8 LOUNGE
9 I.T. / MECH
10 CLEAN SHOP
11 DIRTY SHOP
12 GAME ROOM
13 WAR ROOM
14 BIKE STORAGE
15 BATHROOM
16 KITCHEN
17 CLOSET

One & Co位于旧金山，是日受瞩目的新生代工业设计公司之一。公司成立于10年前，当时选择了教会区一处建于世纪之交的石砖仓库为家。当公司需要向旁扩建时，三位负责人发现他们可以利用这次机会创造一个全新的工作空间，进而体现公司的合作精神、设计价值和不断上涨的名气。

One & Co新工作室604平方米的面积几乎是原工作室的两倍，室内设计的亲密感能够促进员工进行很好的交流。开放式设计利用了很多朝东和朝南的窗户，沿着内墙设置了很多房间——装配室、休息室、“策划室”、厨房和卫生间。设计小组可以临时使用“策划室”，需要时还可以拉上窗帘来保持私密性。

新建筑的极简主义抽象艺术风格为老建筑增添了不少特色：透明和半透明的设计、材料、数量、纹理、色彩和灯光都为人们增添了视觉感受和触觉体验。室内整体的中性色调设计被家具、地板和墙面上的桃红色突显出来，同时桃红色调也激活了人们对深度的感知。胡桃木的建筑板和家具也带来了暖意，自然地将新旧、完美和不完美的世界融合在一起。

Design Company Iris Steinbeck Architekten

COMPANY REPRESENTATION IN BERLIN

Location
Kurfürstendamm 103-104, 10711 Berlin, Germany

Area
1,500m²

Photographer
Werner Huthmacher, Oranienstr. 19a, 10999 Berlin

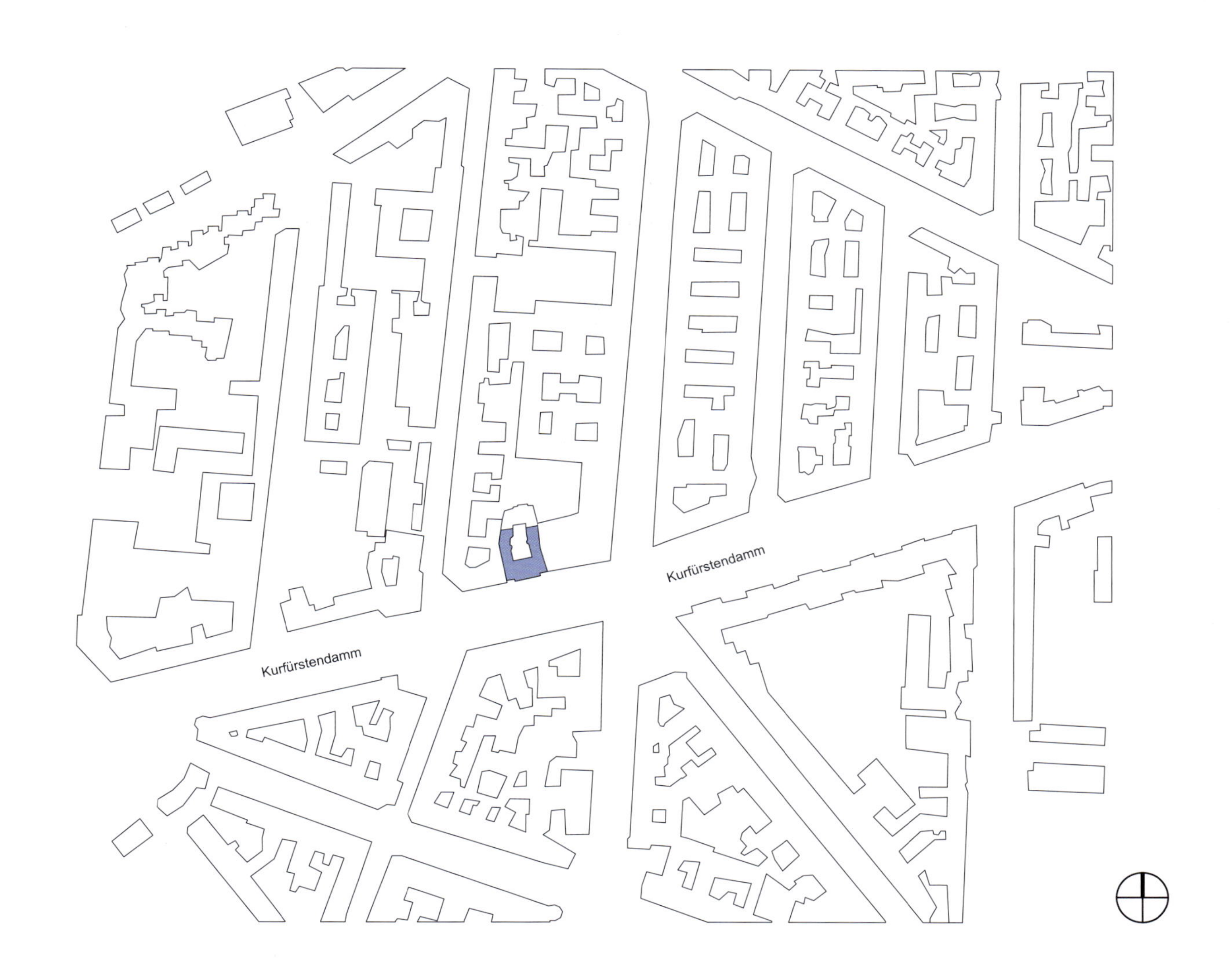

Lageplan M 1:5000

For the design of the Berlin representation of Orco Germany Iris Steinbeck Architekten relied on the dissolution of segmented usage areas. The entry is dominated by an organically shaped reception desk. This element corresponds with a media wall in the waiting area. This wall simultaneously acts as frame for the blue backlit company logo as well as two integrated plasma screens, which can be concealed behind a sliding door if required. The reception area is connected with the open kitchen area, which is used for presentations, working, eating, entertainment and team meetings. An elevated kitchen block serves as a table. Sideways, the monolith is bordered with polished stainless steel, a reference to the desks of the employees. This detail refers to the double use as desk respectively dining table.

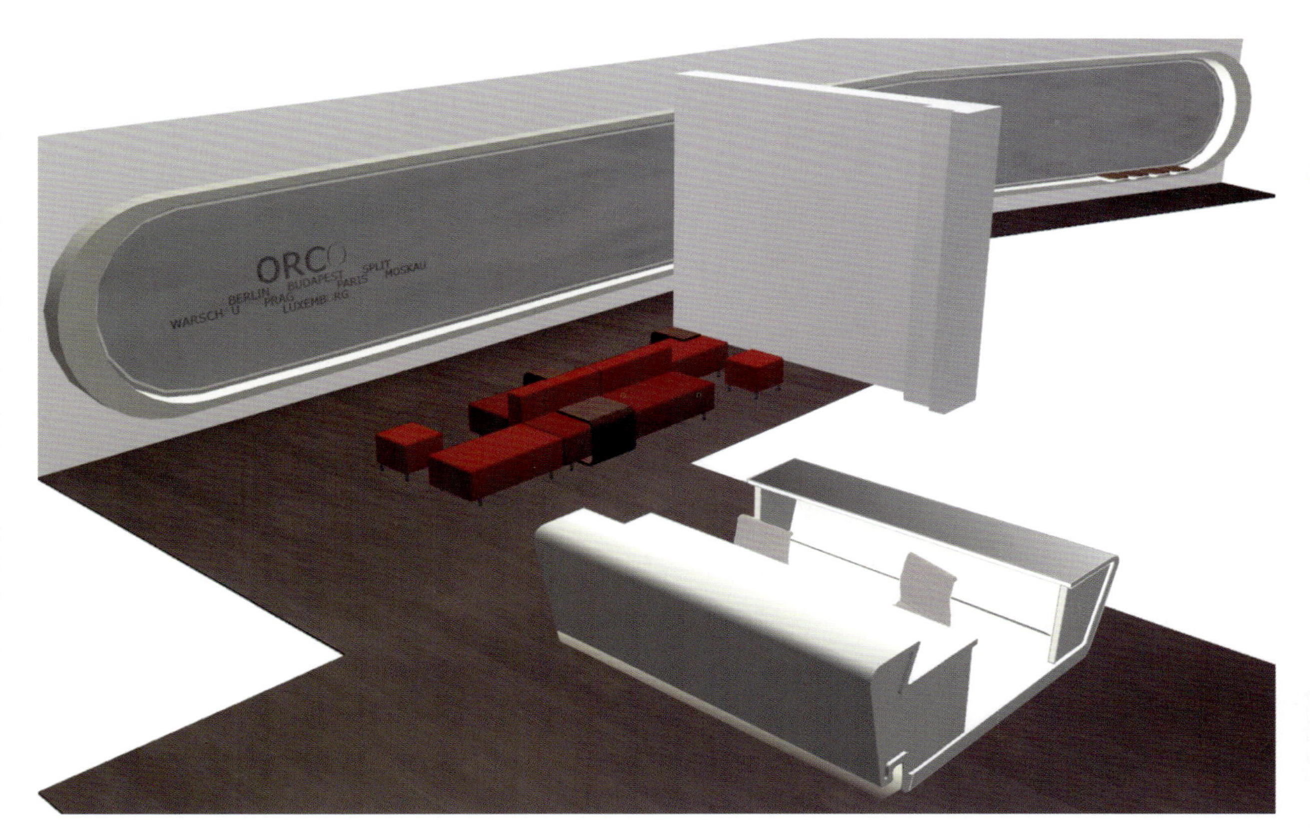

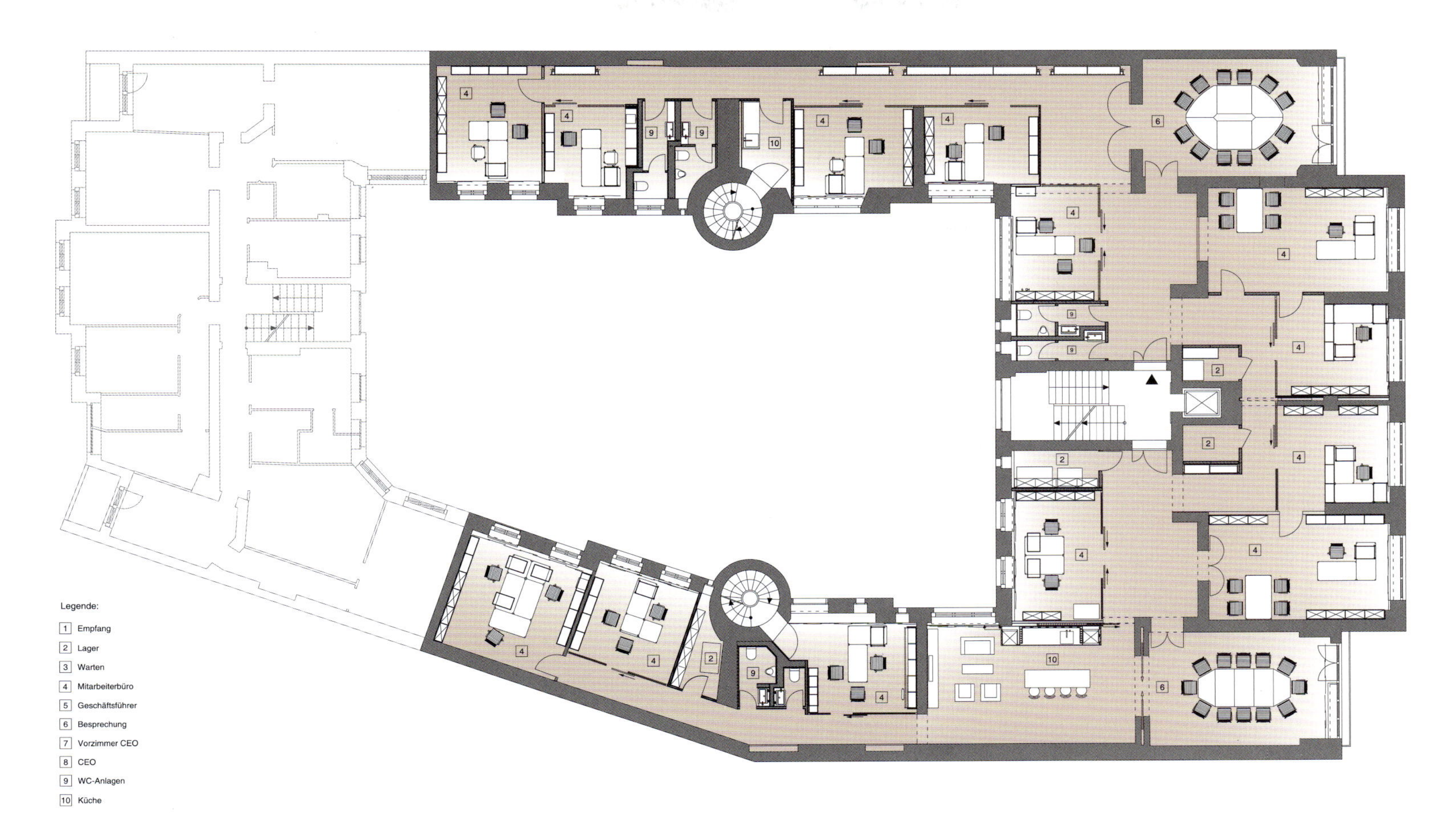

Grundriss 1.Obergeschoss M 1:50

1st floor plan

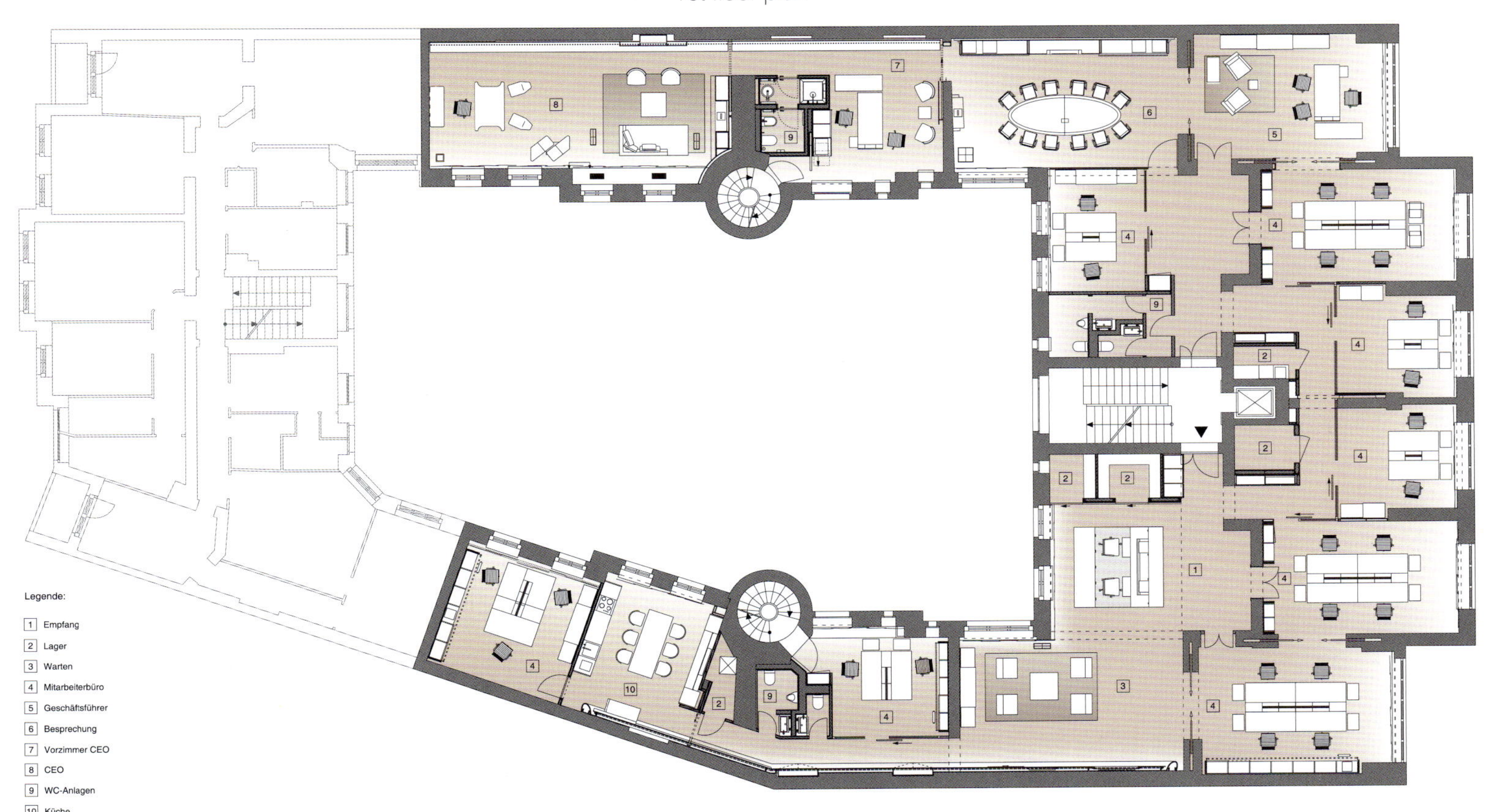

Grundriss 2.Obergeschoss M 1:50

2nd floor plan

为了设计阿科德国在柏林的代理处，艾瑞斯·斯坦贝克建筑事务所采用了分散隔断使用区域的方法。公司的入口是一个有机形状的前台。这个元素与等待区的媒体墙相呼应。墙上还有蓝色背光的公司标志和两块整合的等离子屏幕。这两块屏幕还可以根据需要隐藏到一扇滑动门后面。接待区与开放式厨房相连，在这里可以做报告、工作、就餐、娱乐或召开团队会议。一块凸起的体块充当了桌子。桌子的四周由光滑的不锈钢镶边，与员工办公桌的元素相呼应。这样的设计使这张桌子既可以办公又可以就餐。

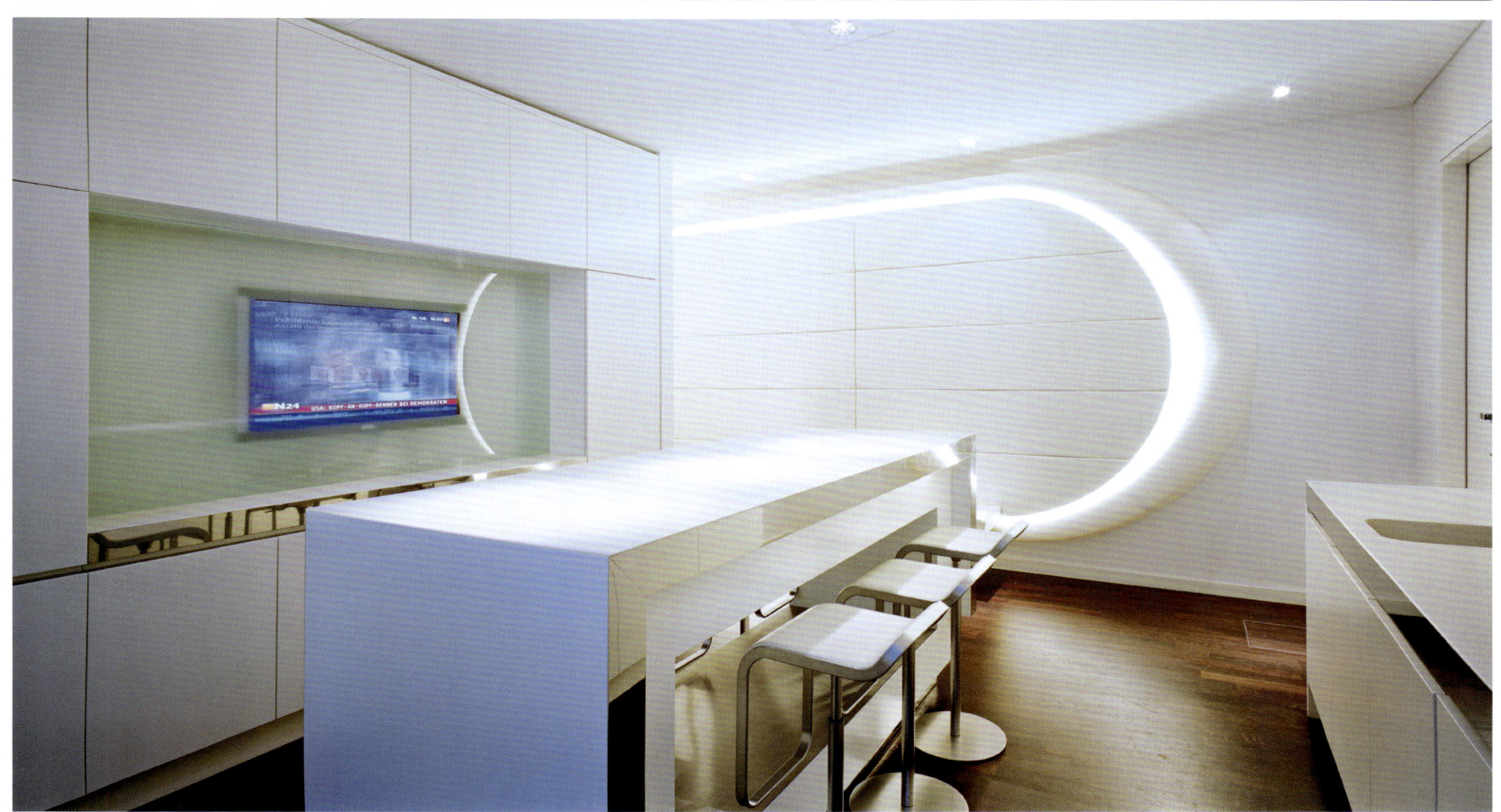
N24
USA: KOPF-AN-KOPF-RENNEN BEI DEMOKRATEN

Designer Maurice Mentjens

POSTPANIC

Design Company
Postpanic

Location
Amsterdam, the Netherlands

Area
565m^2

Photographer
Arjen Schmitz

The Postpanic is divided into two parts: ground floor and first floor.

Ground Floor

The tall hall, a neutral space with bold elements, at the same time functions as the entry and as exhibition space, is in use for seminars and film screenings and acts as the office's living room. Wedged between two columns is a monumental, oak grandstand that takes up a quarter of the studio's width and doubles as stairs to the mezzanine. The grandstand is facing a screen that's suspended above the bar. This detached bar, tiled in white tiles, is simultaneously an autonomous object and recalls an old-fashioned kitchen.

Parallel to the facade, diagonally placed, is a grand table, meant for reading and dining. This 16-seater (5m×1.2m) holds a lowering in its centre to store books and magazines. Bar, grandstand, table and screen together make up the office's "recreation zone". From time to time the employees, sitting on the grandstand, a beer in hand, enjoy a film or football match together.

Attached to the ceiling above the grandstand, an installation of fluorescent tubes radiates over different areas.

Attached to two columns in the kitchen are two wooden beams that both serve as a bookcase and as a demarcation between the kitchen and the adjacent production room.

First Floor

The first floor houses, at both sides of the landing, the staff room and the design studio. Landing and grandstand, both in oak, together form one integral object. The grandstand's steps continue just above the level of the landing, thus forming a cupboard to house the beamer. The balustrade that continues over the entire width visually connects the different areas on the first floor.

Postpanic 分为一层和二层两部分。

一层

一层高大素雅的大厅使用大胆的元素，除了作为出入口和展览空间外，还可作为办公室的客厅使用，讨论会、电影放映也可在这里举行。两根立柱中间立着一个有纪念意义的楔形橡木看台，占据了工作室宽度的 1/4，从这里可至楼梯上的夹层。看台正对面的屏幕悬浮在吧台上方。贴着白色瓷砖的独立吧台同时也是隔断，能使人回想起老式的厨房。

门的斜对面，与之平行的是一个供员工用餐和阅读的大桌子。桌子有 16 个座位（桌子尺寸 5 米 × 1.2 米），中间凹进去的部分可以放书和杂志。吧台、看台、桌子和屏幕一起构成了办公室的“娱乐区”。员工们时不时地坐在看台上，拿着啤酒，欣赏着电影或球赛。

看台上方是连接至天花板的荧光灯管，发出的光可以照射到不同的区域。

连接厨房两根立柱的是两根木梁，木梁既可以当书架用，又是划分厨房和旁边生产室的隔断。

二层

二层空间包括楼梯平台两侧的员工办公室和设计室。橡木制成的楼梯平台和看台合成一个整体。看台的台阶正好延伸到楼梯平台的平面，形成一个能放投影仪的柜子。楼梯的扶手横向延伸，看上去使二层的不同区域连接在一起。

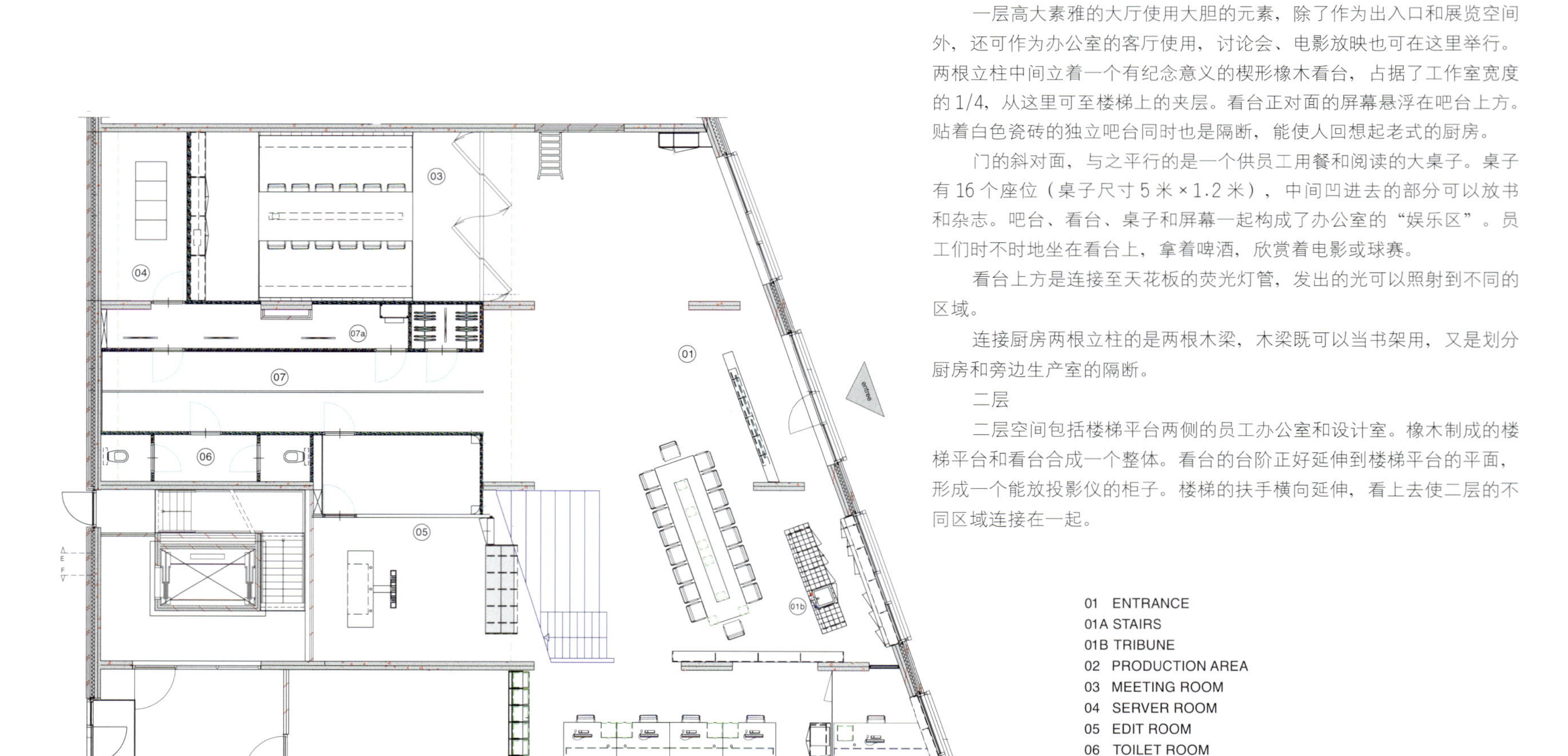

01 ENTRANCE
01A STAIRS
01B TRIBUNE
02 PRODUCTION AREA
03 MEETING ROOM
04 SERVER ROOM
05 EDIT ROOM
06 TOILET ROOM
07 CORRIDOR
07A STORAGE ROOM
08 STUDIO 2D/3D
09 OFFICE MISCHA /JULES
10 TOILET/SHOWER ROOM

Design Company 100% interior Sylvia Leydecker

SAMAS HEADQUARTERS

Project Description
Samas, Worms, Germany

Location
Worms, Germany

Area
900m²

Photographer
Karin Hessmann, Dortmund

A new appearance for the Samas Headquarters, Germany – one of Europe's largest office furniture producers – had become necessary. Samas shall be understood as one big family, welcoming international visitors from all over the world.

First of all, the foyer with reception and lounge-area, open-space-office follows and flows smoothly into each other. The open-minded interior design corresponds to outside nature, as there's a connecting oscillating view axis from entrance through foyer, offices and adjacent garden. Rounded forms of outside garden architecture are integrated into interior design, which generates fluid barriers.

Entering the building, clients feel welcome by a feelgood-lounge, in a well-being warm color-concept. Brownish strip-curtains provide privacy, protect from disturbance, but at the same time allow for watching the surrounding lobby. Poodle carpet and extra-upholstery of series-lounge-furniture provide a quiet atmosphere. Luxury warm golden 3D-carved wall-cladding reinterprets nature's forms.

The same 3D-hightech-surface is found at the reception desk in pure lacquered white. A wall panel describes daily actions happening in office spaces such as "communicate", "develop" and last but not least "dream".

Even washrooms integrate seamless in the concept: 3-Letter-Codes of several international airports, playing with visitor's origins, are found as cabin door-applications. So choose carefully the one to enter.

Open-space-office unites different working-teams. Organic forms are adapted in the carpet flooring and help structure these spaces. A "main road" meanders through the office landscape, providing practicable storage space on both sides.

Meeting space is varying in different types from formal meeting room to informal meeting circle. Working itself tends to be either formal by using separate desks, or even informal and unconventional, working as a team, for instance spontaneously, on a long bench. A fresh and natural landscape for vital office workers.

THINK
denken
VISIONS
visionen
INSPIRATION
CREATIVITY
kreativität
LACHEN
laugh
IMAGINATION
TRÄUMEN
dream
KOMMUNIZIEREN
communicate
TALK
reden
LISTEN
zuhören
ACTIVITY
aktivität

作为欧洲最大的办公家具制造商之一，为了向全世界展现其“大家庭形象”以迎接来自世界各地的客户，Samas 德国总部的重新装潢势在必行。

首先，设有接待区和休闲区的门厅以及开放式办公区交错相通。灵动的室内设计与外部自然环境遥相呼应，一条动态视觉轴线将入口、大厅、办公室和一旁花园的景色连成了一体。户外花园为圆形构造，与室内设计相辉映，形成了一道道流动的屏障。

步入大楼，以暖色调为主题的休息室让客户倍感温馨。褐色的条形窗帘营造出的私密感，让人远离外界纷扰，同时又可以看到大厅里来往的人群。拉毛地毯和精装的办公家具打造出一种静谧的氛围。奢华的金色立体浮雕墙再次诠释了自然的本质。

接待台上也有被漆成白色的立体浮雕。墙板描述了办公区域里每天发生的事：“交流”“发展”以及最重要的“梦想”。

就连洗手间也被天衣无缝地融入到了整个设计理念中：洗手间的门上有若干国际机场的三字代码，客户需要通过组合的代码识别出性别。因此，进去前需小心确认。

开放式办公区集合了不同的工作团队。地面上铺设的地毯具有不同的有机图案，有助于界定各个工作区。一条“主干道”蜿蜒地穿过办公区域，为两旁提供了适用的储藏空间。

会议区分为从正式会议室和非正式会议室，房间的风格千变万化。在这里工作，你可以选用正式会议室里的独立办公桌，也可以选择以团队形式自由办公，比如，自然地围坐在一条长椅上。这里为那些重要身份的员工们创造了清新自然的办公环境。

samas

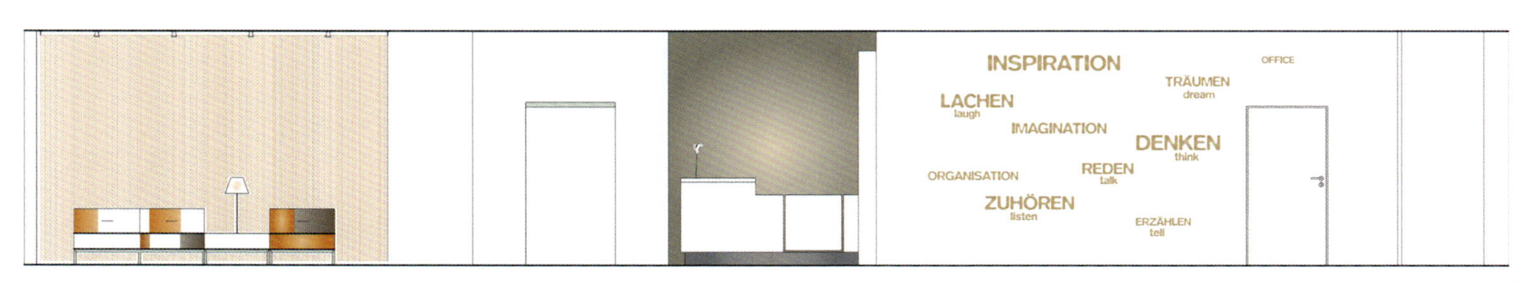
INSPIRATION
OFFICE
TRÄUMEN
dream
LACHEN
laugh
IMAGINATION
DENKEN
think
ORGANISATION
REDEN
talk
ZUHÖREN
listen
ERZÄHLEN
tell

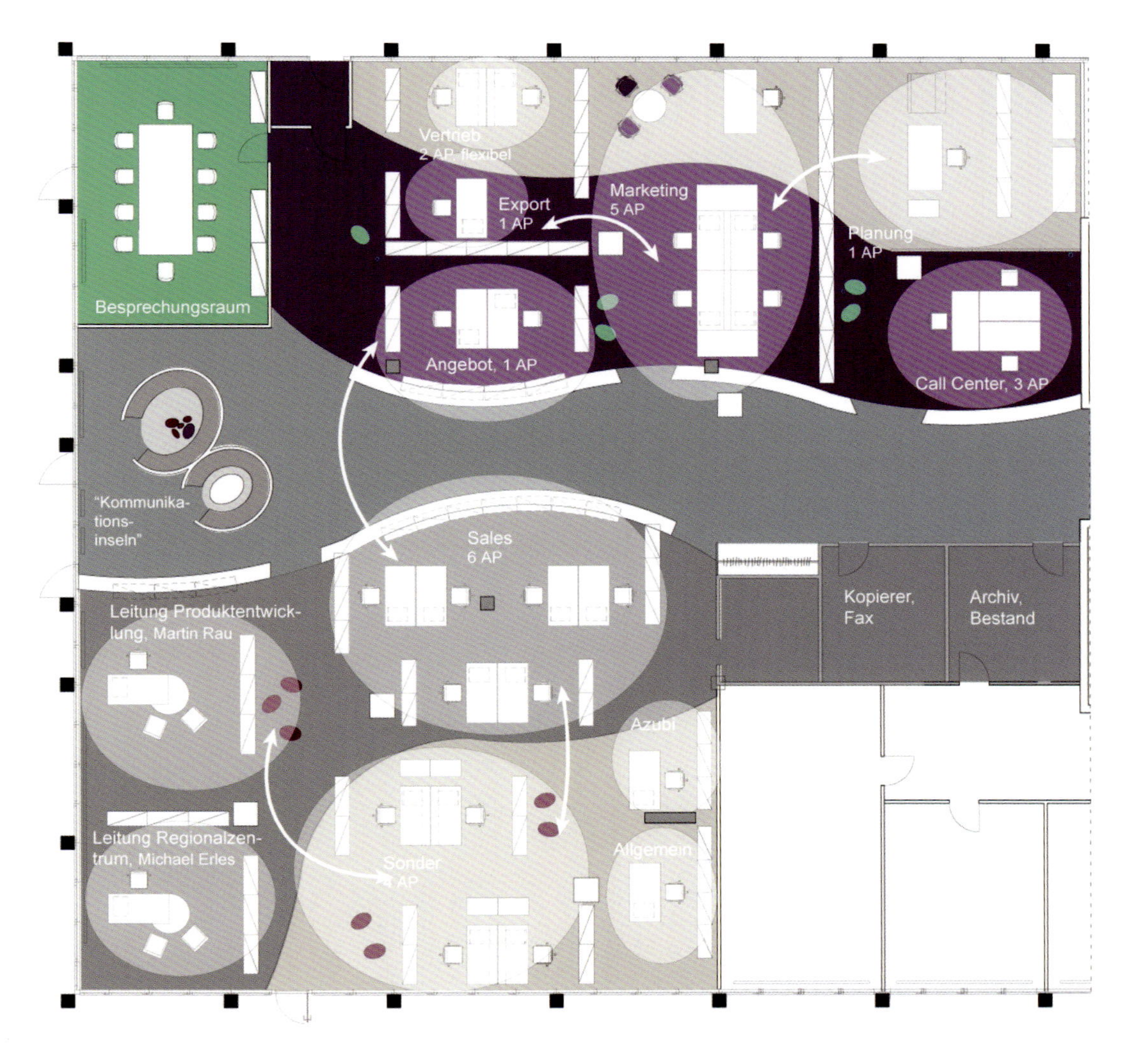

Besprechungsraum
Vertrieb
2 AP flexibel
Export
1 AP
Marketing
5 AP
Planung
1 AP
Angebot, 1 AP
Call Center, 3 AP
"Kommunika-
tions-
inseln"
Sales
6 AP
Kopierer,
Fax
Archiv,
Bestand
Leitung Produktentwick-
lung, Martin Rau
Azubi
Allgemein
Leitung Regionalzen-
trum, Michael Erles
Sonder
4 AP

BREWMATIC 141

LHR

INDEX
索引

DWP | DESIGN WORLDWIDE PARTNERSHIP

dwp | design worldwide partnership is an award-winning, one-stop integrated design service, with global reach. Even in the most challenging of locations, over 450 multi-cultural professionals work together to deliver architecture, interior design, planning consultancy and project management, across borders, to the highest international standards. Divided into three distinct dwp portfolios, namely lifestyle, community and work, dwp integrated design services are managed and driven by the different dwp studios across the globe, ensuring dwp delivers each project with the qualities of a highly focused and specialised service, while offering diversity, flexibility and creativity over a broad spectrum. With currently 12 offices in 10 different countries, dwp presents its finest iconic designs time and again.
www.dwp.com

STUART MARTIN – WAM

Stuart is a founding partner of WAM, an architectural practice based in London, UK.
WAM specializes in high quality sustainable design solutions that deliver award-winning office buildings, interiors, civic buildings, retail outlets, homes and occasionally products.
They are able to provide a comprehensive design service that is creative and forward thinking and deliver flexible dynamic solutions that are respectful of the environment and the planets natural resources.
By promoting the adoption of flexible and collaborative design strategies their office buildings and workplace design interiors have championed a holistic approach to working processes to deliver a dynamic and fertile business environment. This approach has inspired it's users to innovate throughout their working practices and provides a formula for business growth.
www.walkerandmartin.co.uk

STEFAN CAMENZIND

Camenzind Evolution was founded in 1995 from the passion and drive of Stefan Camenzind to create a new level of client-focused design solutions. Over the last twenty years Stefan has successfully innovated a diverse array of international projects, ranging from cultural landmarks and offices to residential and commercial buildings. Stefan leads the project strategy and set-up, which he uniquely tailors for each project. By analyzing clients'objectives and aspirations he evolves a creative process to inspire innovative design solutions.

USOARQUITECTURA

usoarquitectura (use-architecture) is an studio devoted to architecture and interior design. They understand architecture as a whole and that is why they approach projects without rules. Their team is formed by architects, designers, engineers and consultants specialized in all the fields of building. They use the best technology, resources and tools to satisfy their clients' needs. They design spaces with the highest quality, efficiency and integration of architecture to the day to day activities. In every project they achieve the expected results combining experience and innovation, contributing with solutions and taking risks.
As experts in the use of the space their main goal is to maximize its value. With tools such as quick delivery, efficiency and high-tech resources they always search for the best quality in the space requested by the client.
usoarquitectura is the professional and life project of Gabriel Salazar and Fernando Castañón. Since 2005 they have printed their decisive, impeccable quality aesthetic and beyond the line proposals into solutions for their clients to use architecture.

SUSIE SILVERI & ANNE-MARIE CHARLEBOIS

Anne-Marie Charlebois
Over the last 14 years, Anne-Marie (Senior Designer) has gained valuable experience in designing projects from small to large scale, both as a project coordinator and a senior designer. She understands the importance of comprehensive design processes that result in successful projects.
Excelling in developing innovative design elements, she has both organizational and interpersonal skills that play a vital role in seamlessly transitioning all the phases throughout a project.

Susie Silveri
For the past 28 years, she has gained recognition in her field becoming a specialist in design of corporate offices, with extensive experience in the implementation of law firm projects.
Her acute sense of scale, ability to visualize space, and her problem-solving capabilities result in aesthetic and functional spaces. Her intuitive response to information from her clients instils the spirit of her projects.
Susie directs, coordinates and supervises all aspects related to the realization of the projects. She successfully brings forth her concepts and her experience and ensures completion of the mandate by a close, collaborative effort with the client.

BBFL

Broad-Based, Fore&Liberate
Where interaction and communication plays the key role in the design process. On point with the clients' needs and wants, they act on an individual or group to condition behavioral patterns within the work environment. One needs to understand the client's office culture, behavior and needs. Only then, through the designing process, issues and solutions are being put forward to by both parties. From acknowledging the situation, regardless of one's requirement and behavior, creative solutions are implemented through the crafting and styling at the space.
Whether it means to improve one's productivity or to kick a bad habit in the office.
Aside from the usual aesthetic of a space, the fore front of design conceptualizing has to be constantly push in this current ever changing concrete jungle they live in. Moving away from the tradition and the customary practice, will allow ample room for higher quality staff work flow and more conductive working environment.
Their design of intent and fundamental practice, is strongly based on this basic ideolory that aid in reaching higher standard of environment they work in. Improving their corporate life style and increasing productivity.

IPPOLITO FLEITZ GROUP

Ippolito Fleitz Group was found in 2002. Managing Partners are Gunter Fleitz & Peter Ippolito.
Gunter Fleitz:
Study of architecture in Stuttgart, Zürich and Bordeaux
worked with Steidle+Partner, München
Project management for the Federal Supreme Court Leipzig for Prof. Stübler
1999 Founding member of zipherspaceworks
Member of Bund Deutscher Architekten BDA
Peter Ippolito:
Study of architecture in Stuttgart and Chicago
worked with Studio Daniel Libeskind, Berlin
Assistant to Prof. Ben Nicholson, Chicago
1999 Founder member of zipherspaceworks
2001/02 Visiting professor at the Academy of Fine Arts Stuttgart
2004~2008 teaching position at the University of Stuttgart
2009 teaching position at the University of Biberach

IRIS STEINBECK ARCHITEKTEN

Iris Steinbeck
born in Frankurt / Main
1981-1986 studied Architecture in Darmstadt, Germany
1986-1991 Architect in Los Angeles, California, USA
1991-1995 developed and managed the Berlin branch of a Frankfurt based architectural firm
1995 established Iris Steinbeck Architekten
1998 founding director and member of the Executive Board of KW Friends, KW Institute for Contemporary Art , Berlin

CARY BERNSTEIN ARCHITECT

The office of Cary Bernstein Architect is committed to progressive design resulting from thoughtful planning, focused attention to detail and the highest construction standards. Each project is developed in response to the unique combination of client, site and budget without the imposition of a preconceived style or solution. In addition to the promotion of architectural excellence, they offer exceptional service to their clients through all phases of design and construction.
Cary Bernstein graduated from Dartmouth College in 1984 with a B.A. in Philosophy and Russian Literature. She received an M.Arch. from the Yale School of Architecture in 1988.
Cary practiced in New York for 6 years prior to opening her San Francisco office in 1995. The firm's projects have won numerous design awards. Cary has taught both the Russian language and jewelry fabrication at Dartmouth College, philosophy at Yale University and architectural design at U.C. Berkeley. She is currently an Adjunct Professor of Architecture at the California College of the Arts and serves as Chair of the SFMOMA A + D Forum. Cary is a member of the 1% Solution, providing a minimum of 1% of her firm's working hours to assist non-profit organizations with their design and planning needs.

ENOTA

Enota was founded in 1998 by Aljoša Dekleva, Dean Lah and Milan Tomac with the ambition to create contemporary and critical architectural practice of an open type based on collective approach to development of architectural and urban solutions. Over the years Enota has been constantly developing and from the beginnings it has served as creative platform for more than fifty architects. Since 2002 Enota's partner architects are Dean Lah and Milan Tomac.
Dean Lah was born in 1971 in Maribor, Slovenia and graduated from the Ljubljana Faculty of Architecture in 1998. In the same year he cofounded the architectural office Enota where he works as partner architect ever since.
Milan Tomac was born in 1970 in Piran, Slovenia and he graduated from the Ljubljana Faculty of Architecture in 1998. In the same year he cofounded the architectural office Enota where he works as partner architect ever since.

GONZALO MARDONES VIVIANI

Gonzalo Mardones Viviani was born in Santiago de Chile on July 8th 1955. He gets his degree as architect from the Universidad Católica de Chile, where he graduates with the Maximum Honors. He receives the First Prize in the Architecture Biennale, for the best degree project among all the Architectural Schools in Chile, for his project for urban renewal of the South-West Center of Santiago. He has been a professor of architectural design workshops and directed degree projects in the Faculties of Architecture of the Universidad Católica, Universidad de Chile, Universidad Central, Universidad Andrés Bello and Universidad Finis Terrae, in addition to having been guest professor and lecturer in different universities in Chile, and abroad. His work has been published by the main architectural magazines and honored at Biennales. He has been a member of the National Commission of Competitions of the Architects Association in Chile and a Founding member of the Association of Architectural Practices (AOA).

I29 | INTERIOR ARCHITECTS

i29 I interior architects, a creative and versatile interior design studio,aims at creating intelligent designs and striking images. Space is the leitmotiv, the result always clear, with a keen eye for detail. They try to get to the core of things but keep it looking simple.
They have been nominated and won several awards like the Rotterdam Design Award, Dutch design awards, LAI awards and The Great Indoors award. They won the Dutch Design Prize for best interior design and The Great Indoors Award for best office design.

HOFMAN DUJARDIN ARCHITECTS

Hofman Dujardin Architects was founded in 1999. Since then, they have been working on a wide variety of architectural, interior and product design projects with a team consisting of approximately eight architects. This wide scope of projects has been a well considered decision. The diversity in design issues entails an enrichment of any project. Their team switches easily from innovative product development to e.g. pragmatic dwelling floorplans.
Their clients mainly are developers, housing corporations, multi-nationals, law firms and private clients.
Their main goal is to create inspiring buildings, interiors and products that enhance life at large. A surrounding in which people can live and work in an optimized way and where the investments made are fully effected .
For office buildings they design creative working environments in which the employees can work highly efficient and they are inspired continuously. Condominiums and dwellings they transform into interiors with maximum comfort and fully adjusted to ideas and wishes of the clients.

ONG&ONG PTE LTD

With a track record of almost 40 years in the industry, ONG&ONG has earned an unparalleled reputation for integrating skilled architecture, clever interior design, creative environmental branding and sensitive landscape design. Paramount to their success is their insistence on servicing their clients with creativity, excellence and commitment. They continually strive to uphold their mission to be the designer of they age – a premier design practice both locally and in the region.
In addition to projects in Singapore, ONG&ONG has also completed large-scale developments regionally. This has prompted the setting up of offices in China, Vietnam, India and Malaysia. In-depth knowledge of the local context, culture and regulations allow them to better understand their clients' needs. They are an ISO14001 certified practice and consistently strive to meet and exceed their clients' expectations. To grow their international reputation, ONG&ONG has now set up an office in New York, USA.

NANCY KEATINGE & STANLEY FELDERMAN

Born in California, Nancy Keatinge graduated from the University of Southern California and received her Masters Degree from Northwestern University.
Ms. Keatinge's work has been widely recognized with a number of awards, including the "Best Executive Office Design" from Interiors for Columbia Pictures, "Restaurant of the Year" from Restaurant and Hotel Design magazine for the Le Triangle restaurant, an Honor Award from the Los Angeles chapter of the American Institute of Architects for the CalMat corporation headquarters and an IIDA award for White O'Connor Curry & Avanzado.

Born in New York City, Stanley Felderman graduated from the city's High School of Music and Art with high honors and then from Pratt Institute in Brooklyn with a degree in architecture.
Stanley's work has been widely recognized with a number of awards, including the "Best Executive Office Design" from Interiors for Columbia Pictures, "Restaurant of the Year" from Restaurant and Hotel Design magazine for the Le Triangle restaurant, an Honor Award from the Los Angeles chapter of the American Institute of Architects for the CalMat corporation headquarters, and the RAS Industrial Design Award for Mr. Felderman's design of a G-2 jet for the L.J. Hooker Corporation. Mr. Felderman has judged design competitions worldwide, and has won numerous design awards.

GERD PRIEBE

Everything started with Gerd Priebe's grandfather who was a master of the building trade. His assiduous and precise work fascinated his grandson, filling him with enthusiasm. With his kind-heartedness and sternness, Priebe's grandfather conveyed respect for the precise execution of all craftwork. He also guided Priebe's curiosity at an early age in a direction which he was to continue by taking up his training as a carpenter and which he was to complete as an architect. During his studies in Wuppertal, his thirst for knowledge led Priebe to Erich Schneider-Wessling and Gottfried Böhm. After completing his studies in Wuppertal as a graduated architect, Priebe continued to study under Joachim Schürmann and Oswald Matthias Ungers. This is where he learned to appreciate the significance of space, proportions, light, plasticity and functions.
Priebe's great role model however was Richard Meier whom he met during his studies. Following study trips to France, Turkey and Italy, Priebe visited Richard Meier in New York. It was a time which left its mark on Priebe who returned to Germany in 1991 with a wealth of experience and started to work for the Hungarian architect Pal Dévény for four years. During this time he visited Dresden in 1995, and it was love at first sight. Another three years were to pass before he moved to Dresden and founded Priebe Architektur. His office has since successfully participated in numerous national and international competitions, and it now possesses a commendable client base.

ROBERT MAJKUT

Robert Majkut is one of the most important European designers. His hardearned brand and consistency in his approach to design makes him recognizable as a very popular designer and creator of unique places.
Born in Szczecin, Poland in 1971, he graduated with honours from an art secondary school with furniture design as major. Having first studied architecture at Szczecin University of Technology, he then moved to the Academy of Fine Arts in Poznan. Finally he graduated from Cultural Studies at Poznan University.
Robert's work received many accolades – in 2002 he was honoured with the "Rising Star" award by the British Council. He was also nominated for Elle Style Awards 2007 in the "good design" category. In 2011 Polish marketing magazine Brief included him in the ranking of 50 most creative people in business.
He has been active in the field of popularising good design. Member of the programme council of New Culture Bec Zmiana Foundation; guest speaker at universities, judge at competitions, expert quoted in Polish and foreign press.

TAYLOR SMYTH ARCHITECTS

At Taylor Smyth Architects they look to create "poetry" in each project. And they find inspiration, drawn or interpreted, from both the unique vision and identity of their clients and the environments in which they live, work and play. They embrace certain fundamental elements to which they believe all people respond: daylight, texture and colour, natural materials and the symbiotic relationship between space and architecture. They believe that attention to detail and the creative use of materials add a distinctive and memorable experience to every project.
They are dedicated to excellence and the creation of inspired environments – spaces that delight, uplift and engage.
Their experience ranges from large, complex educational and community environments, to commercial and workplace projects and custom residential design. Each project features a unique blend of innovation and technical know–how.
Their work has received extensive recognition through design awards and publications both nationally and internationally.

SUB ESTUDIO

Architecture and Design office formed in 2009 by Isabel Nassif, Julia Masagão and Renata Pedrosa.

SPACES ARCHITECTS

Spaces Architects was established in March 1999 with a mission of creating sensible, functional spaces enhanced by the intangible sense of emotion, of power, of playfulness. This results in architecture that can be extraordinary responding to your unique needs throughout the design process, is just as much an architect's mission as shaping aesthetically inspired built environment through communication.
The team is led by Kapil Aggarwal and Nikhil Kant passed in year 1996 from Manipal Institute of Technology, India.
Their motto is to design, detail and enjoy working in responsive to their client's need in very honest, efficient and professional manner. They believe that their works follow a unique style based on their ideology and perspective towards the design requirement.

DAVID GIANOTTEN

David Gianotten joined OMA in 2008 and became partner in charge of OMA Asia in 2010. In 2009, he launched OMA's office in Hong Kong, China which is now OMA's Asia headquarters. He oversees OMA Beijing next to OMA Hong Kong and leads OMA's development in Greater China and the rest of Asia. Projects currently under his supervision include the Shenzhen Stock Exchange, the Taipei Performing Arts Center, and the Chu Hai College of Higher Education in Hong Kong. In 2010, he delivered the West Kowloon Cultural District conceptual masterplan, the biggest cultural project in Hong Kong to date, together with Rem Koolhaas. Born in 1974 in the Netherlands, he studied Architecture and Construction Technology at the Eindhoven University of Technology. Before joining OMA, he was the Managing Director – Architect of SeARCH.

PIERANDREIASSOCIATI
BEHAVIORAL DESIGNERS & ARCHITECTS

PIERANDREI ASSOCIATI

"We see our profession as based in a culture of making seeking the perfect balance between the infinite components of the design process: science and art, modernity and tradition, historical memory and a clear vision of the future".This sentence represents at the best the philosophy of Pierandrei Associati, an international fluid-thinker design group.
In 2007 and 2010 Pierandrei Associati was selected for the "New Italian Design" exhibition at the Triennale di Milano and still in 2010 PA has been selected by the ADI Index in Italy and won the Red Dot Award (Germany) and the Best Design Award (USA), for the best product with "Beta", office furniture system designed for Tecno. Recently PA also won the US Award 2010 for the best interior design.

JOSE ABEIJÓN VELA & MIGUEL FERNÁNDEZ CARREIRAS

Jose Abeijón Vela
Architect_ Etsac, Universidade Da Coruña
Colegiated in 1996_ Nº 2059 Belonging to Coag, from Spain

Miguel Fernández Carreiras
Architect_ Etsac, Universidade Da Coruña
Colegiated in 1997_ Nº 2171 Belonging to Coag, from Spain

Location
C/ Juan Flórez 118, 1º; 15005 A Coruña, Spain
Telephone: +34 981 153 544 // +34 615 994 965
Fax: +34 881 925 229
Web: www.abeijón-fernández.com
E-mail: jabeijón@abeijón-fernández.com
miguelfer@abeijón-fernández.com

DIPEN GADA & ASSOCIATES

Dipen Gada & Associates, which is popularly known as DGA, began as a very modest Architectural & interior design firm. Gradually with time and every project accomplished, DGA evolved from an exclusive interior design firm to a civil and architectural planning firm and attained the position as one of the reputed firms of Baroda.
The principal designer and founder of the firm, Dipen Gada holds a Bachelors degree in Civil Engineering from M.S University.
The core team at DGA consists of qualified and driven professionals comprising of Architects, Interior Designers and Engineers who create versatile body of work ranging from architecture, interior, landscape and product design. The firm strives to maintain a balance between aesthetics and functionality in all its designs.

ZANE TETERE

- Award "Best private interior design in Latvia 2009" for private house on Zaru street, Riga
- Award "Arhip – The architecture Award 2009-Innovation" in Moscow
- Winner in international competition "Harbour bench" in festival "Baltic Breeze"
- Award BEST OFFICE AWARDS 2011 in Moscow as best foreign interior for McCann-Erickson Riga and Inspired office
- Award "Best office interior 2011 in Latvia" for McCann-Erickson Riga and Inspired office

MOSSINE & PARTNERS

Mossine & partners architectural studio was founded in 1998 in Berlin. The Moscow office was opened in 2007. Main activities are:
-urban planning,
-residential and civic buildings design
-interior design
Mossine & partners is the winner of the 13th international festival of architecture and design 2011 (Moscow) and of the 11th international festival of architecture and design 2009 (Moscow).
In 2011 Mossine & partners was nominated to the LEAF Awards (London) and Best Building Awards (Moscow).

SYLVIA LEYDECKER

Sylvia Leydecker, Dipl.-Ing., is interior architect BDIA and director of the Cologne based studio 100% interior. She studied interior architecture in Wiesbaden and Jakarta and today designs Corporate Interiors for businesses. Her references include amongst others Bitkom, Evonik and Samas. Recently she was awarded several times for product design as well. Sylvia Leydecker is the author of, *Nanomaterials in Architecture, Interior Architecture and Design* published by Birkhauser. Sylvia is a frequent lecturer on nanotechnology and interior architecture; she has published numerous articles on the subject.

PEDRA SILVA ARCHITECTS

Pedra Silva Architects is an architectural design office based in Brighton and Lisbon catering for all aspects of architectural and interior design. Their work covers small-scale projects with an emphasis in leading retail brands to large scale multi-million investments.
In retail they develop architectural designs associated with pioneering concepts for restaurants, fashion and clothing, health and office spaces. They currently work with a portfolio of international clients that are key players in these sectors.
On larger scale projects, they develop all phases of project design including project management.Besides architectural services their offices provide a hub that bridges cultural gaps for foreign investment in local economies.
Majority of work is currently in Europe with current expansion moving towards emerging markets in Africa, Middle East and Asia.

HEAD ARCHITECTURE AND DESIGN LTD.

Head Architecture and Design Limited was established in Hong Kong,China by a group of Architects, designers and project managers who shared the common goal of the pursuit of excellence in architectural design. The scope of their experience broadly covers all aspects of building projects from inception and budget establishment, through brief development, conceptual and developed design on a wide range of architectural and interior projects.

During their careers Head Architecture and Design Limited staff have been extensively involved in projects of varying size and complexity and in many regions throughout the world, including Europe, South East Asia, the Middle East, New Zealand, in addition to the People's Republic of China. Their multi-disciplinary team enables a comprehensive design service from initial conceptual planning through to supervision and completion of project site works.

MAURICE MENTJENS DESIGN

Maurice Mentjens primarily designs interiors and related objects and furnishings. Creations are almost exclusively for the project sector: shops, hotels and restaurants, offices and museums. Intriguing, smaller scale projects are preferred to larger projects. The aim of this compact, talented and dynamic team is to deliver high-end design reflecting its passion.

Quality and creativity are prioritised in all aspects of the design and implementation process. The agency is three times winner of the Dutch Design Award.

In 2007, the design agency received the Design Award of the Federal Republic of Germany. Clients include the Bonnefantenmuseum Maastricht, Frans Hals Museum Haarlem, Post Panic video producers Amsterdam, DSM Headquarters in Heerlen and Amsterdam Airport Schiphol.

ROSAN BOSCH LTD.

Rosan Bosch Ltd. is a team of artists, architects, designers and academics who work with architectural projects, design, communication and education, process management and process implementation as well as user introduction. Furthermore, they collaborate with a number of partners, specialists and suppliers.

For each project a creative taskforce is gathered who follow the project through all its phases.

They work in all phases of the project; from development of the concept to design and budget proposals to construction management. They are used to working with large scale project as well as minor projects, both in respect to areas, organisation and budgets, and they have extensive experience in firm budget control and research into sponsors and other financing of projects.

ARCO ARQUITECTURA CONTEMPORÁNEA

ARCO Arquitectura Contemporánea was founded by architects Bernardo Lew and José Lew which at the same time are in charge of the direction of the company. The main activities of the firm are: planning, architectonic project execution, interior design, consulting and coordination of structural projects, MEP systems, site direction, technical and economical management, coordination and construction supervision.

ARCO partners are both architects graduated from Universidad Anáhuac and have more than 15 years of professional experience en México and the US. Their projects have been published in national and international magazines and have received important awards. They are constantly invited to give lectures in congresses and architecture schools.

ENRICO TARANTA

The internationally-oriented team is led by Enrico Taranta. Before opening his own office in 2007 in Shanghai, he worked at Massilimiliano Fuksas and SMC Alsop. For Massilimiliano Fuksas, Taranta was involved in the design development of the Rome congress center "the cloud", which almost finished construction. For SMC Alsop he worked on the design of the Shanghai "Bei Wai Tan" international Cruise terminal. He has been collaborating with designers from Copenhagen, New York, Miami, Shanghai and Nigeria. He collaborated with PTW Architects for the design of the "Water-cube" Beijing Olympic swimming center. At a wide range of universities around China Enrico Taranta has been invited to give a lecture and he was a permanent lecturer at the interior design faculty of Donghua University RDI Shanghai. Enrico Taranta himself received his master diploma at the faculty of architecture in Rome.

mapos llc

MAPOS

Their name, "Mapos", is a combination of "map" – meaning to map, strategize, and to help people find their way – and "os" which stands for Open Source, a free sharing process where everyone's input is valuable and essential. They are an integrated team of architects, designers, and hard-to-label creative instigators who work with their clients and colleagues on creative concepts, architectural environments, and experiential branding. Recognized as a leader in sustainable design, Mapos creates customized design processes that incorporate community, reuse, technology, and great design.